Biotransformation of Organic Nitrogen Compounds

Progress in Basic and Clinical Pharmacology

Vol. 1

Series Editors
P. Lomax, Los Angeles, Calif.
E.S. Vesell, Hershey, Pa.

KARGER

Basel · München · Paris · London · New York · New Delhi · Singapore · Tokyo · Sydney

Biotransformation of Organic Nitrogen Compounds

Volume Editors
A.K. Cho, Los Angeles, Calif.
B. Lindeke, Solna

29 figures and 11 tables, 1988

KARGER

Basel · München · Paris · London · New York · New Delhi · Singapore · Tokyo · Sydney

Progress in Basic and Clinical Pharmacology

Cover illustration
Computer-generated model of an artificial protein showing hydrophobic contacts (blue van der Waals surfaces of all leucine side chains) in the interior of a template-assembled 4 helix-bundle topology (TASP). (According to M. Mutter et al., in PEPTIDES, Proc. 10th Amer. Peptide Symp., 1987, in press.)

Library of Congress Cataloging in Publication Data
Biotransformation of organic nitrogen compounds
Volume editors, A.K. Cho, B. Lindeke.
(Progress in basic and clinical pharmacology; vol. 1)
Includes bibliographies and index.
ISBN 3-8055-4650-5
1. Organonitrogen compounds – Metabolism. 2. Drugs – Metabolism. 3. Biotransformation (Metabolism) 4. Cytochrome P-450. I. Cho, A. K. (Arthur K.), 1927 – . II. Lindeke, B. (Björn). III. Series.
[DNLM: 1. Drugs – metabolism. 2. Nitrogen – metabolism. 3. Oxidation – Reduction. QU 55 B616)]
QP801.N55B563 1988
615'.314-dc 19
DNLM/DLC
for Library of Congress 87-33863

Drug Dosage

The authors and the publisher have exerted every effort to ensure that drug selection and dosage set forth in this text are in accord with current recommendations and practice at the time of publication. However, in view of ongoing research, changes in government regulations, and the constant flow of information relating to drug therapy and drug reactions, the reader is urged to check the package insert for each drug for any change in indications and dosage and for added warnings and precautions. This is particularly important when the recommended agent is a new and/or infrequently employed drug.

Printed in Switzerland by Druckerei 'Der Bund', Bern
ISBN 3-8055-4650-5

Contents

IV. Nitrogen Oxygenation by Flavin Enzymes

V. Nitrogen Oxidation in Carcinogenesis

VI. Nitrogen Oxidation in Toxicity

VII. Metabolic Disposition of Nitrogen Functionalities

Preface

The extensive use of nitrogen containing organic compounds, as pharmaceuticals and chemical intermediates, exposes man to a large number of these compounds under widely varying conditions. Before their elimination, these compounds are usually metabolized by the oxidative enzymes commonly involved in xenobiotic biotransformation. This monograph reviews the biochemistry of these biotranformation reactions and their involvement in the toxicity and elimination of organic nitrogen compounds. This area of pharmacology and toxicology is attracting investigators with widely differing research interests and, by summarizing relevant information into a single volume, we hope to broaden the perspective of interested individuals.

Prog. bas. clin. Pharmac., vol. 1, pp. 1–5 (Karger, Basel 1988)

Introduction

Arthur K. Cho

Department of Pharmacology, UCLA School of Medicine,
Center for the Health Sciences, Los Angeles, Calif., USA

Nitrogen-containing substances are the most common of all of the organic compounds of pharmacological interest. Pharmacological or therapeutic interest in nitrogen-containing compounds dates back to the use of alkaloids or extracts containing them for diverse effects, such as stimulants (cocaine, nicotine, caffeine), analgesic/euphoriants (morphine), tranquilizers (reserpine) and decongestants (ephedrine). Despite their prevalence and the extensive information on their pharmacology, little was known of the metabolic changes in the nitrogen functionality of these compounds until the mid-1950s. Thus, although the occurrence of metabolic nitrogen-oxidation was indicated early [5], the major pathways of metabolism were up to the mid-1950s considered to be accountable by oxidative or hydrolytic reactions at nearby carbon atoms, and the role of nitrogen in metabolism was considered less important.

The impetus for investigation of nitrogen metabolism came from a toxicological interest in organic compounds that was developing in the 1960s. Oxidation of nitrogen was shown to be the activating step for the carcinogenic aromatic amines and for hemolysis associated with aromatic amines. These observations led to a major effort to understand the enzymology [6, 17] and chemistry [2, 10] of nitrogen oxidation. As evidence for recent interest, three international symposia have been organized to focus exclusively on the metabolism of nitrogen functionalities [1, 3, 4] since 1970. The purpose of this monograph is to bring together what the editors think are the important aspects of nitrogen oxidation that relate to the pharmacology and toxicology of organonitrogen compounds.

Nitrogen functionalities exhibit a wide range of chemical reactivities over the multiple oxidation states in which they can exist. Recognition of these different oxidation states and the varied chemical reactivities associated with each of them is vital to an understanding of organic nitrogen metabolism. For example, the properties of the amine state are dramatically altered when it is protonated. Since the enzymes involved in amine oxidation are membrane-bound, the less polar, free base form is the likely substrate. However, the often quoted higher rates of metabolism associated with increasing lipophilicity of amines could be due to their greater partition into the lipid matrix of the enzyme – to provide a higher local substrate concentration – and have little to do with their affinity or efficiency as substrates. Higher oxidation states of nitrogen have also been shown to be biologically important.

The nitrogen-hydroxy state is the active precursor function in aromatic amine carcinogenesis [12] and aniline induced hemolytic anemia [9]. Compounds containing the NO function interact with the heme moiety in hemoglobin and cytochrome P450 to impair their normal function. The reduction of the nitro group in nitrofurans and nitroimidazoles has been shown to generate a reactive electrophilic product. This product forms covalent bonds with proteins and other macromolecules that are thought to contribute to the toxicity of these nitro compounds [7].

Nitric oxide, generated from nitrate or nitrite hydrolysis, has been shown to activate guanylate cyclase and generate cyclic guanosine-3′,5′-monophosphate, which causes smooth muscle relaxation [8]. Studies with aliphatic nitro compounds have shown that these substances undergo cytochrome P450-mediated denitrification generating nitrite and a carbonyl compound [16]. If this reaction occurs in liver cells, the resulting activation of guanylate cyclase should be considered a possible pharmacological effect of denitrification. Denitrification is also thought to contribute to the toxic reactions of nitrosamines.

The availability of a convenient and simple assay for formaldehyde has permitted elucidation of fundamental mechanisms of cytochrome P450-mediated oxidation at carbon atoms α to the nitrogen (nitrogen-demethylation) [11], but studies on nitrogen oxidation mechanisms have been slower because of the complexity of the reactions and the multiplicity of products. More recent progress in this area of research has utilized specific analytical procedures and the tools of mechanistic chemistry, such as stable isotopes and electron spin resonance. The use of gas chromatography (GC) and high-pressure liquid chromatography (HPLC) with different detectors has allowed the identification and quantification of specific reaction products, some of which were unstable. For example, a specific and sensitive GC assay for aminofluorene nitrogen-hydroxylation showed that this reaction could be increased threefold by selectively inhibiting another cytochrome P450 reaction [14]. This observation has serious implications for changes in carcinogenicity of this aromatic amine. The development of HPLC allowed the separation and quantitation of nonvolatile, thermally unstable compounds. Compounds containing oxidized states of nitrogen are frequently unstable, so that HPLC with electrochemical detection is an excellent method of analysis. Indeed, the formation of *N*-acetyl-*p*-quinoneimine in liver microsomes was demonstrated by HPLC separation and electrochemical detection [13]. The authentic material was electrochemically generated and the identification of microsome derived quinoneimine was demonstrated by the electrochemical behavior of this highly reactive species. Free radicals or functionalities with an unshared electron have been proposed as intermediates in several toxic reactions involving nitrogen functions. Their presence in incubation mixtures of tissue components has been demonstrated by electron spin resonance techniques to have half-lives sufficient to achieve measurable steady-state levels [15]. Thus, with different analytical techniques, mechanisms for the metabolism of organonitrogen compounds are being elucidated.

This monograph is organized into chemical, biochemical, toxicological and pharmacological sections. In chapter I, Jon M. Fukuto and Arthur K. Cho present a selected review of nitrogen functional group chemistry emphasizing those reactions that are related to its biochemistry. A major focus of the chapter is on the oxidation reactions of amines and hydroxylamines because of their frequent occurrence in metabolic reactions.

The next three chapters cover the biochemistry of nitrogen oxidation. Chapters II and III review the interactions of nitrogen functions with cytochrome P450. First, Gerald T. Miwa and John S. Walsh describe the biochemistry of this heme system and its demonstrated role in oxidation and reduction reactions. Then Bjorn Lindeke and Ulla Paulsen-Sörman discuss the chemistry and biochemistry of complexes formed by oxidized intermediates of nitrogen functions. Russell A. Prough presents a discussion of the flavin monooxygenase system, the first enzyme system to be shown unequivocally to N-oxidize xenobiotics, in chapter IV.

The next two chapters summarize the undesirable effects organic nitrogen functions have on the biological system. In chapter V, Check Y. Quon reviews the carcinogenicity of these compounds and in chapter VI Russell A. Prough and Philip Wiebkin describe their toxicity. In the final chapter, Arthur K. Cho summarizes the metabolic reactions of these functional groups. Although oxidation is the most common initial step, the subsequent reactions, such as conjugation, often determine the ultimate fate of nitrogen functions.

It is apparent that the area of organic nitrogen biotransformation is rapidly developing with respect to mechanisms and specific reactions. The compilation of the available information during a rapid growth phase may be a foolhardy task since the resulting document could be obsolete before it is published. The authors were all aware of this problem and have attempted, whenever possible, to discuss and present information in mechanistic, chemical terms so that the information can be related to newer relevant findings. It is the opinion of the editors that the organization of the information on the basis of the chemistry of the compounds rather than their pharmacology or toxicology is a more logical format for this reason.

References

1 Bridges, J.W.; Gorrod, J.W.; Parke, D.V.: Biological oxidation of nitrogen in organic molecules (Taylor & Francis, London 1972).

2 Damani, L.A.: Oxidation at nitrogen centers; in Jakoby, Bend, Caldwell, Metabolic basis of detoxication. Metabolism of functional groups, pp. 127–149 (Academic Press, New York 1982).

3 Gorrod, J.W. (ed.): Biological oxidation of nitrogen (Elsevier/North-Holland, Amsterdam 1978).

4 Gorrod, J.W.; Damani, L.A.: Biological oxidation of nitrogen in organic molecules. Chemistry, toxicology and pharmacology, p. 443 (Horwood, Chichester 1985).

5 Heubner, W.: Studien über Methämoglobinbildung. Arch. exp. Path. Pharmak. *72:* 241–281 (1913).

6 Hlavica, P.: Biological oxidation of nitrogen in organic compounds and disposition of N-oxygenated products. CRC crit. Rev. Biochem. *12:* 39–101 (1982).

7 Holtzman, J.L.: Role of reactive oxygen and metabolite binding in toxicity. Life Sci. *30:* 1–9 (1982).

8 Ignarro, L.J.; Lippton, J.C.; Edwards, J.C.; Baricos, W.H.; Hyman, A.L.; Kadowitz, P.J.; Gruetter, C.A.: Mechanism of vascular smooth muscle relaxation by organic nitrates, nitrites, nitroprusside and nitric oxide: evidence for the involvement of S-nitrosothiols as active intermediates. J. Pharmac. exp. Ther. *218:* 739–749 (1981).
9 Kiese, M.: The biochemical production of ferrihemoglobin-forming derivatives from aromatic amines, and mechanisms of ferrihemoglobin formation. Pharmacol. Rev. *18:* 1091–1161 (1966).
10 Lindeke, B.: The non- and postenzymatic chemistry of N-oxygenated molecules. Drug Metab. Rev. *13:* 71–121 (1982).
11 Lindeke, B.; Cho, A.K.: N-Dealkylation and deamination; in Jakoby, Bend, Caldwell, Metabolic basis of detoxification, pp. 105–126 (Academic Press, New York 1982).
12 Miller, E.C.; Miller, J.A.: Mechanisms of chemical carcinogenesis: nature of proximate carcinogens and interactions with macromolecules. Pharmacol. Rev. *18:* 805–838 (1966).
13 Miner, D.J.; Kissinger, P.T.: Evidence for the involvement of *N*-acetyl-*p*-quinoneimine in acetaminophen metabolism. Biochem. Pharmacol. *28:* 3285–3290 (1979).
14 Razzoak, C.; Mercier, M.; Roberfroid, M.: Induction, activation, and inhibition of hamster and rat liver microsomal arylamide and arylamine N-hydroxylase. Cancer Res. *40:* 3540–3546 (1980).
15 Stier, A.; Clauss, R.; Lucke, A.; Reitz, I.: Redox cycle of stable mixed nitroxides formed from carcinogenic aromatic amines. Xenobiotica *10:* 661–673 (1980).
16 Ullrich, V.; Hermann, G.; Neber, P.: Nitrite formation from 2-nitropropane by a microsomal monooxygenase. Biochem. Pharmacol. *27:* 2301–2304 (1978).
17 Ziegler, D.M.: Microsomal flavin-containing monooxygenase: oxygenation of nucleophilic nitrogen and sulfur compounds; in Jakoby, Enzymatic basis of detoxication, vol. 1, pp. 201–227 (Academic Press, New York 1980).

A.K. Cho, PhD, UCLA Medical Center, Department of Pharmacology,
23–272 Center for the Health Sciences, 10833 Le Conte Avenue,
Los Angeles, CA 90024–1735 (USA)

Prog. bas. clin. Pharmac., vol. 1, pp. 6–26 (Karger, Basel 1988)

I. Chemistry of Organic Nitrogen Compounds

Arthur K. Cho, Jon M. Fukuto

Department of Pharmacology, UCLA School of Medicine,
Center for the Health Sciences, Los Angeles, Calif., USA

Introduction

This chapter summarizes chemical properties of the different oxidation states of organic nitrogen compounds that are relevant to their biochemistry. The objective is to sensitize the reader to the relationship between the chemistry of organic nitrogen and its metabolic disposition. The different nitrogen functional groups shown in table I and their chemical properties are discussed in increasing order of oxidation state. Detailed presentations of organic nitrogen chemistry may be found in reference sources listed in the bibliography of this chapter [25, 54, 64]. The chemistry of the nitrogen functions involves an unshared or lone pair of electrons that participates in protonation, oxidation and reduction reactions to varying degrees depending on the oxidation state of the nitrogen and on the organic radical attached to it. The enzymatic transformations in which these functions participate have chemical counterparts and an understanding of them should clarify the biochemical processes.

Amines

Chemical Properties

The amino state is the most common oxidation state of bioactive molecules. An amine exists in aqueous solution as an equilibrium

Table I.

Structure		Name
R-NH_2		amine
↓		
R-NH-OH		hydroxylamine
│	R-NH_2-NH_2	hydrazine
│	$R_3N \rightarrow O$	N-oxide
↓		
R-N-O·		nitroxide
↓		
R-N=O		nitroso
	H R-C=NOH	oxime
	R-C=N-R ↓ O	Nitrone
	R-N=NH	diazene
	R-N=N-R ↓	azo
	R-N=N-R ↓ O	azoxy
↓		
RNO_2		nitro

mixture of the neutral, free base form and the ionic, ammonium ion form:

$$H^{\oplus} + NR_3 \rightleftarrows HNR_3^{\oplus}$$

This equilibrium is expressed by the pK_a, or the negative logarithm of the dissociation constant of the ammonium ion or conjugate acid, so that the larger the pK_a, the stronger the base. Since there is a large difference in the chemical properties of these two forms, the base strength of these functions has an enormous influence on amine biochemistry. The ionic species is much more polar or hydrophilic and tends to be water-soluble with a very low organic/aqueous partition coefficient relative to the free base. This property of amines is used in pH partition processes where substances are separated on the basis of differences in their pK_a

and their partition coefficients [18]. In the procedure, a pH is chosen that separates amines from metabolites by the difference in the partition of the neutral species. Charged ammonium ions can also be extracted into organic solvents by the use of lipophilic counter ions, such as picrates or sulfonates, to form lipophilic ion pairs which readily extract into organic solvents [63]. This feature is utilized in the extraction of polar amines and quaternary ions from aqueous solutions.

The difference in physical and chemical properties between the ammonium ion and the free base contributes to differences in their pharmacological properties. For example, studies of tertiary amine cholinergic agonists have shown that only the ammonium ion is responsible for receptor activation [9, 35]. Therefore, for these compounds, the proportion of the ionic species is a pharmacological determinant. In contrast, the free base form of tertiary amines is important in their interaction with cytochrome P450 [11] and monoamine oxidase [53], so the apparent rates of enzymatic reaction vary with pH in the opposite direction. Knowledge of the base strength and extent of protonation is, therefore, essential for interpreting the interaction of organic bases with biological systems.

Structural Effects on Base Strength. The pK_a values of organic bases vary with structure and in some cases it is possible to predict, with reasonable certainty, the pK_a of a particular compound using data available in tabulated form in reference books. These predictions are based on the effects that structural changes have on the availability of the electron pair for interaction with protons. Table II demonstrates the effects that attached groups have on the base strength of amines. The reduction in base strength associated with structural changes of ethylamine or aniline result from changes in electron distribution within the molecule and a reduction in the availability of the unshared electron pair on nitrogen for interaction with acids. Note, for example, the profound effect of the cyano group on pK_a values. There are two factors which, in many cases, account for the effects of different groups (substituents) on base strength, induction and resonance. Induction refers to the asymmetric distribution of electrons as a result of dipole interactions, and resonance refers to the delocalization of electrons that result from the presence of unsaturated functions in the molecule. Resonance effects are much more pronounced in aromatic systems and pK_a values are accordingly more sensitive to substitution patterns.

Table II. Base strength of amines [from ref. 1]

		pK_a
Aliphatic amines		
CH_3H_2-NH_2	ethylamine	10.63
$NCCH_2$-CH_2-NH_2	2-cyano-	7.70
$HOCH_2$-CH_2-NH_2	2-hydroxy-	9.50
CH_3OCH_2-NH_2	2-methoxy-	9.45
$PhCH_2$-CH_2-NH_2	2-phenyl-	9.83
Aromatic amines		
Aniline		4.50
4-Nitroaniline		1.02
4-Cyano-		1.74
4-Methyl-		5.12
4-Methoxy		5.29

The changes in base strength that result from substitution changes in closely related compounds can be predicted or calculated using equations and constants that have been empirically determined. This approach is based on a linear free energy relationship that is assumed to exist between the physical properties of closely related compounds so that the logarithms of their rate or equilibrium constants are related by a linear equation reflecting the substituent and the reaction or equilibrium considered. This relationship was first proposed in quantitative terms by Hammet [33] who defined a reaction constant rho (ρ) and a substituent constant sigma (σ), such that:

$$\text{Log } K_s/\log K_o = \rho \times \Sigma\sigma,$$

where K_s and K_o are the equilibrium or rate constants for the substituted and unsubstituted reactions, respectively, and Σ is the sum of all substituent constants. The relationship for the dissociation of anilinium ions is:

$$\text{Log } K_a'/\log K_a = 2.77 \times \Sigma\sigma,$$

where K_a' and K_a are the dissociation constants of the conjugate acids of the substituted and unsubstituted aniline, respectively.

Sigma values have been determined empirically for a large numer of substituent groups from dissociation constants for substituted benzoic acids and tables of such values, identified by position on the aromatic ring, have been published [42]. Subsequently, analogous equations and constants were developed for aliphatic systems [68] which have been used to predict base strength [12, 31]. Details of the calculations and their accuracy have been discussed in a number of monographs that also list the substituent constants [1, 33, 37, 58].

The substituent effects reflected by sigma values are due to inductive or resonance contributions to base strength but there are steric contributions to base strength because of the solvation of protons. The access of such solvated protons is limited when bulky groups surround the nitrogen. For example, Brown and Kanner [8] showed that 2,6-di-t-butylpyridine ($pK_a = 3.58$) was a much weaker base than 2,6-dimethylpyridine ($pK_a = 5.77$) although there was very little difference in the electronic effects of the two alkyl radicals.

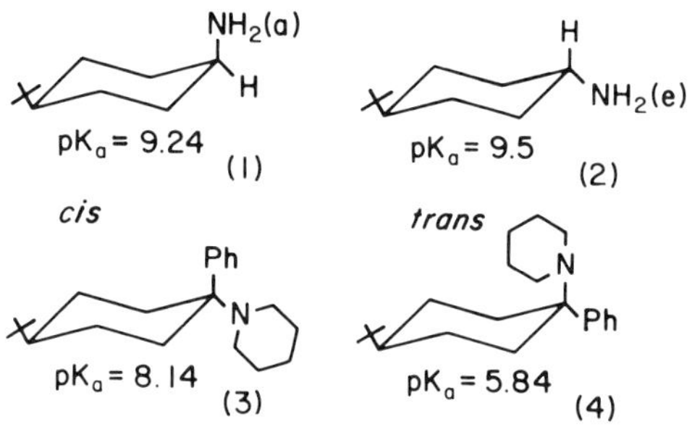

The equatorial and axial conformations of the amino group in cyclohexane derivatives provide another example of steric hindrance to ionization. In a study of the pK_a values of cis (structure 1) and trans (structure 2) 4-t-butylcyclohexylamine, Tichý et al. [70] have shown that the cyclohexyl conformer with the equatorial amino function is a stronger base. The 4-t-butyl group is fixed in the equatorial position on the ring because of its bulk and in turn fixes the conformation of the other groups attached to the cyclohexane ring. Thus, cis-4-t-butylcyclohexylamine has the amino group in the axial position and the trans-isomer in the equatorial, or less hindered, position. A more exaggerated example of this phenomenon is provided by the work of Geneste et al. [29] who examined the pK_a values of two analogues of phencyclidine (structures 3, 4).

Again, the isomer with the equatorial nitrogen is the stronger base and the difference in pK_a between the equatorial (structure 3) and axial (structure 4) amines is two orders of magnitude. This difference was interpreted in terms of the more hindered axial nitrogen being more difficult to protonate. The two phencyclidine derivatives also displayed marked differences in their octanol-water partition coefficients with the axial isomer favoring octanol by more than four orders of magnitude [44]. The differences presumably are due to greater hindrance to solvation of the axial conformer such that it is essentially a hydrocarbon.

Reactions with Carbonyl Groups. Primary and secondary *amines* can react with carbonyl functions to form carbinolamines which dehydrate to imines. Secondary amines form the positively charged iminium ion that can lose a proton to become an enamine [19] (scheme 1):

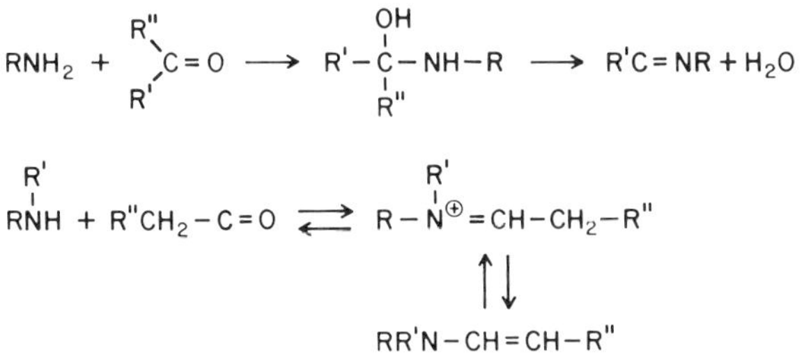

Since the reaction is reversible, imines or immonium ions can be converted back to the starting amine and carbonyl compound in aqueous solution. These adducts are the initial step in physiological transamination reactions catalyzed by pyridoxal containing enzymes where the pyridoxal carbonyl condenses with an amino group. Carbinolamines have been proposed as intermediates that subsequently break down to the carbonyl compound and amine in the N-dealkylation reaction catalyzed by cytochrome P450. An enamine was found in the gas chromatographic effluent of a mixture of phencyclidine metabolites [32]. However, the compound may not be a true metabolite since it could arise as an artifact of the assay system. Phencyclidine is a tertiary amine so elimination of water would generate either an iminium ion or an enamine. Only the enamine would be volatile and detectable by the gas chromatographic technique used. The iminium ion is also formed in incubation mixtures of liver preparations [71] and is thought to be the species responsible for inhibition of cytochrome P450 by covalently binding to it [39, 56].

The major differences between aromatic and aliphatic amines are due to the delocalization or resonance interaction of the lone pair with the pi electron system of the aromatic ring. This delocalization reduces base strength and nucleophilicity in aromatic amines (pK_a 4–5) compared to aliphatic amines (pK_a 9–10). The aromatic ring also provides an electron sink that can delocalize a charge, thereby allowing other parts of the molecule to participate in reactions originating on the nitrogen. For example, the ethanol metabolite, acetaldehyde, could condense with the amino group of dopamine to form an imine which then reacts with the aromatic ring to form an isoquinoline derivative. This product has been proposed as a participant in the chronic pharmacology of ethanol [13, 16] (structure 5):

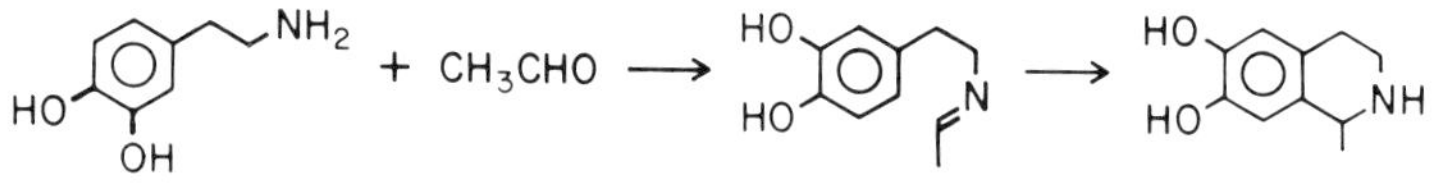

(5)

Oxidation

The chemical oxidation of amines can occur by at least four mechanistically distinct pathways. One involves a sequential electron transfer from the amine to the oxidant during which an amine radical cation is formed. Another pathway is the direct oxidation of nitrogen by electrophilic oxygen giving, in most cases, hydroxylamine as the intermediate which in turn can be rapidly oxidized. A third possibility is α-hydrogen abstraction by the oxidant leading to α-carbon rather than direct nitrogen oxidation. Finally, hydrogen atom abstraction from the nitrogen can occur to generate a nitrogen centered radical.

These mechanisms have already been proposed for the biochemical oxidation of amines. For example, electron transfer mechanisms appear to be relevant for both the cytochrome P450 and the flavin-containing enzymes. Lindeke and Cho [48] proposed α-hydroxylation and direct two-electron N-oxidation as possible mechanisms for N-dealkylation. Both pathways require formation of a carbinolamine which decomposes to the corresponding carbonyl compound and the dealkylated nitrogen species. The pathway involved in a particular metabolic reaction obviously depends on the enzyme and the chemical nature of the substrate. However, as mechanistic details for the different enzymes are elucidated, more accurate predictions of the metabolic disposition of organic compounds can be made on the basis of the chemical properties of the substrate (see chapter II).

One-Electron Oxidation. Electrochemical oxidations are of mechanistic interest because they proceed by the sequential transfer of electrons from the nitrogen to the electrode and resemble some enzymatic reactions. This oxidation procedure generates radicals as transient intermediates and the products isolated reflect their chemical properties and those of the solvent in which the reaction takes place. Early electrochemical studies performed by Barnes and Mann [5] in nonaqueous media with platinum electrodes revealed that the electrochemical oxidation of primary aliphatic amines follows two pathways, a low-potential pathway and a high-potential pathway, shown in scheme 2:

$$R{-}CH_2{-}NH_2 \xrightarrow{-e^-} R{-}CH_2{-}\overset{\bullet\oplus}{N}H_2 \qquad (1)$$

$$R{-}CH_2{-}\overset{\bullet\oplus}{N}H_2 \xrightarrow[-H^{\oplus}]{} R{-}\dot{C}H{-}NH_2 \qquad (2)$$

$$R{-}\dot{C}H{-}NH_2 \xrightarrow[-e]{} R{-}CH{=}\overset{\oplus}{N}H_2 \qquad (3)$$

$$R{-}CH{=}\overset{\oplus}{N}H_2 \xrightarrow[-H^{\oplus}]{+H_2O} R{-}CHO + NH_3 \qquad (4)$$

$$R{-}CH_2{-}\overset{\bullet\oplus}{N}H_2 \longrightarrow R{-}\overset{\oplus}{C}H_2 + \dot{N}H_2 \qquad (5)$$

$$R{-}\overset{\oplus}{C}H_2 \longrightarrow \text{hydrogen} + \text{polymers} + \text{alcohol} \qquad (6)$$

Equations 1–4 comprise the high potential pathway that requires two electrons and yields ammonia and the corresponding aldehyde. Although acetonitrile was the solvent, sufficient water was present to permit hydrolysis. When this anodic oxidation is performed in aqueous alkali, the increased basicity causes further oxidation to nitriles [34]. The low potential pathway, equations 1, 5 and 6 involves an initial single electron transfer with subsequent carbonium ion formation. t-Butylamine follows the low potential pathway because it has no α-hydrogens and cannot form an iminium ion. The primary products from anodic oxidation of t-butylamine, isobutylene and t-butylalcohol, are the result of carbonium ion formation.

In general, secondary and tertiary amines undergo the same electrochemical oxidation reactions as primary amines [55] but differences exist which could be relevant to biochemical reactions. Dapo and Mann [15] and Russell [61] independently found that the product of anodic oxidation of triethylamine in acetonitrile and dimethylsulfoxide was the protonated triethylammonium ion, i.e. instead of a net oxidation reaction, the amine was converted to its protonated form. An initial one-electron oxidation of the amine which was followed by hydrogen atom abstraction from solvent was proposed to account for the observation but since products that would reflect solvent radical formation were not identified, the proposal has not been evaluated.

When the nitrogen is substituted with allylic or benzylic groups, alkyl cleavage occurs with oxidation, as predicted by the radical in equation 2 (scheme 2). Allyl and benzyl groups form particularly stable radicals whose generation is therefore energetically favored [47, 55]. This generalization is not universal, however, since ethyldiisopropylamine, which could yield the stable isopropyl radical, is cleaved at the ethyl-carbon-nitrogen bond at a rate much greater than expected on the basis of either relative radical stability or random cleavage [65]. This inconsistency was interpreted in terms of a different mechanistic pathway involving an intermediary enamine preferentially formed at the ethyl group.

The oxidation of aromatic amines is more complicated as coupling and hydrolysis reactions can occur [55]. In nonaqueous or acidic conditions, the coupling reactions of aniline during electrochemical oxidation lead to polymerization reactions that generate complicated polyaromatic dyes. In basic solutions, *p*-benzoquinone, which results from the oxidation of the presumed initial product, *p*-aminophenol, can be isolated. If the aromatic ring contains a suitable leaving group, such as chloride, methoxy, ethoxy, etc., coupled products in which a nitrogen moiety is substituted for the leaving group are formed. Azobenzene can also form under appropriate conditions [72].

It should be recognized that the products obtained in these electrochemical studies may not necessarily be identical to products obtained from biochemical systems, even if the biochemical oxidation occurs by a radical cation pathway. Factors such as reaction media (pH, effective water concentration, etc.) and substrate concentration can dramatically affect both the type and relative quantity of product formed. However, for the reasons stated earlier, a superficial understanding of the nature and reactivity of radical cation intermediates generated by electrochemical oxidations should help to interpret amine biotransformation and provide a basis for prediction of metabolic pathways.

Amine radical cations can also be formed by chemical systems. For example, reduction of tertiary amine oxides with iron (II) salts [23, 24, 49] generates radical cations that undergo fragmentation reactions similar to those observed in electrochemical studies. One interesting observation made by Lindsay-Smith et al. [50] was the formation of hydroxy derivatives, such as $R_2N(CH_2)_nCH(OH)Me$ and $R_2N(CH_2)_nCH_2OH$, by oxidation of the N-oxide of R_2NBu (R = Me,Bu) by Fe(II). Presumably, these products form by an intramolecular hydrogen atom transfer of the type depicted in scheme 3:

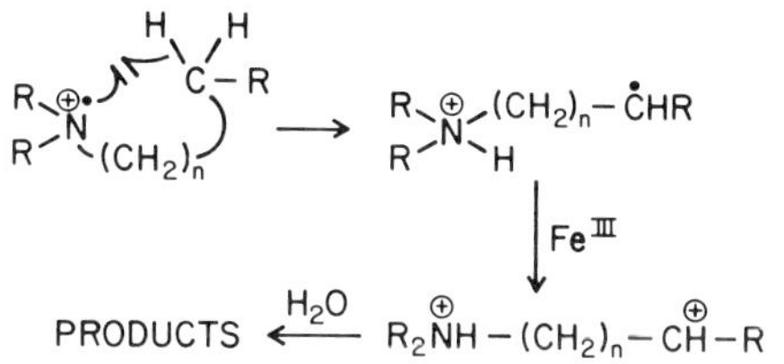

Biochemical Significance of One-Electron Oxidations. As mentioned earlier, the major amine oxidizing enzymes, cytochrome P450 and the flavin containing mono-oxygenase, are capable of performing one-electron oxidations and evidence for involvement of amine radical cations in the oxidative process is accumulating. The oxidizing species in reactions catalyzed by cytochrome P450 is thought to be a high oxidation state iron-oxo complex. When this highly electrophilic species encounters a potential substrate, it would be expected to react preferentially at centers with high electron density, such as the lone pair of electrons on nitrogen. One possible reaction pathway is a direct, one-electron transfer from nitrogen to the electrophilic enzyme. This subject is discussed in chapter II.

Concerted Two-Electron Oxidation. The lone electron pair on nitrogen can also react directly with an electron-deficient atom, such as a hydroperoxide oxygen. This reaction is mechanistically distinct from one electron oxidations and can yield different products. Primary amines are oxidized by peroxidic reagents to the corresponding hydroxylamines which quickly react further to yield oximes, nitroso compounds and eventually hydroxamic acids or nitro derivatives [3, 4]. Tertiary alkylamines that lack α-hydrogens, commonly yield the nitro derivative as the primary product.

Primary aromatic amines are all oxidized to either their nitroso or nitro derivatives by peroxides [10, 20, 21, 59, 73]. This transformation proceeds through the hydroxylamine which further oxidizes. Mechanistic studies on the oxidative pathway indicate that the electrophilic peroxidic oxygen initially attacks the nucleophilic nitrogen lone pair [30, 41]. Secondary amines undergo oxidation to hydroxylamines which also react further to yield nitroxides and immonium radicals, and tertiary amines are oxidized to N-oxides and then to nitrones [46]. Homolytic or electron transfer pathways have been implicated in oxidations of secondary amines with diacylperoxides [40] so that other pathways can compete to a small extent with electrophilic oxygen attack at nitrogen.

Hydrogen Atom Abstraction Oxidation. The oxidation of both primary and secondary amines by *MnO_2*, under either acidic or neutral conditions, involves α-hydrogen abstraction which eventually leads to the dealkylation of nitrogen [23, 24]. This reaction is an example of an oxidation on carbon that is facilitated by the unshared electrons on nitrogen:

$$CH_3CH_2-NH_2 \xrightarrow{MnO_2} CH_3CHO + NH_2$$

The oxidation of primary aromatic amines by MnO_2 gives the coupled azo compounds. The reaction could involve either sequential or concerted transfer of electrons to MnO_2 followed by proton loss and coupling of nitrogen centers, but direct hydrogen atom abstraction from nitrogen by MnO_2 is also possible. Oxidation of groups α- to nitrogen have also been demonstrated in the cyclization of a series of N-substituted ortho-phenylenediamines to the corresponding benzimidazoles by MnO_2 [22]:

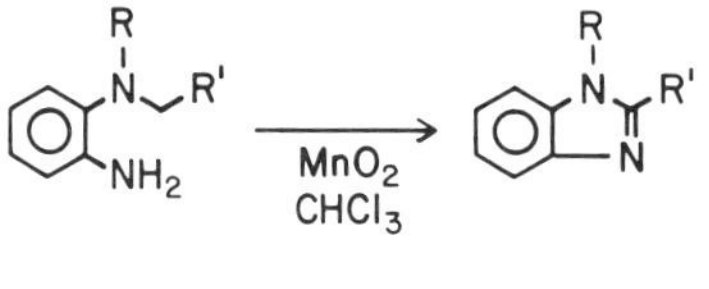

Other nonperoxidic oxidizing agents such as lead tetraacetate or sodium borate will also oxidize primary aromatic amines to the corresponding azo compounds. These reactions could proceed by either direct hydrogen abstraction or electron transfer followed by proton loss. Both pathways generate an aniline radical that couples to form the precursor to the azo derivatives.

Hydroxylamines

Chemical Properties

Hydroxylamines are much weaker bases (pK_a 3–6) than their amine counterparts because of the replacement of H by a hydroxyl group. Organic hydroxylamines can be acylated at the nitrogen or oxygen and the intraconversion of the N-acyl derivative, called a hydroxamic acid to the O-acyl derivative (the acyloxyamine), is thought to be an intermediate step in the chemical carcinogenesis caused by aromatic amines [45]:

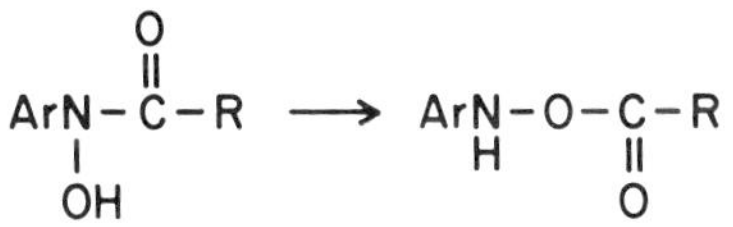

One reaction of mechanistic interest to aromatic amine metabolism is the rearrangement of phenylhydroxylamine to p-aminophenol [54]. This reaction occurs in strong acid and is thought to occur by the loss of water from the protonated hydroxylamine:

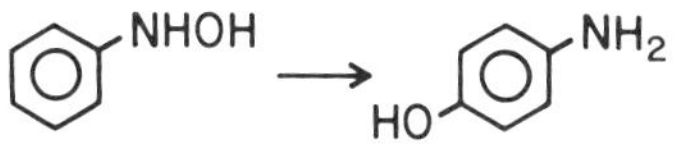

The ortho substituted compounds can also form. This reaction provides a chemical basis for the ring hydroxylation of aniline through N-hydroxylation instead of direct C-hydroxylation.

Oxidation

Hydroxylamines are readily oxidized to nitroso or nitroxide states. The nitroxide function results from a one-electron oxidation of the hydroxylamine state that can be effected by weak oxidizing agents, such as superoxide ion. This oxidation has been shown for the ESR probe

TEMPO, a secondary hydroxylamine [60]. The oxidation of N-hydroxy-phentermine to the corresponding nitroso compound by superoxide has been proposed to proceed through the nitroxide on the basis of a study of the reaction in organic and aqueous media [28]. It is also possible to oxidize hydroxylamines directly to the nitroso state by stronger reagents, such as chromic acid.

Oxidation of *arylhydroxamic* acids with silver oxide generates a series of compounds resulting from the intermediate nitroxide [27]. The aryl acyl nitroxide can acylate the starting hydroxamic acid to form the O,N-diacyl derivative or it can react with aryl nitroso compound generated from the first reaction in an oxygen transfer reaction, the result of which is a disproportionation to the acylamino compound and aryl nitro derivative (scheme 4):

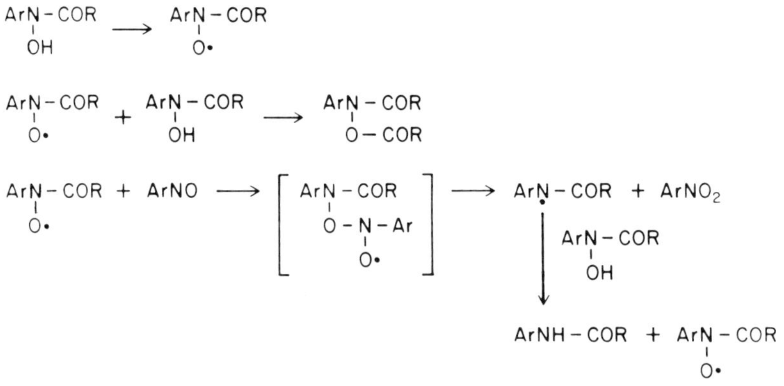

These reactions provide a chemical basis for the O,N-transacylation, the reduction and the oxidation of arylhydroxamic acids in which the initial step is an oxidation of the nitrogen function.

Nitroxides

Chemical Properties

The nitroxide state, a radical formed by a one-electron oxidation of the corresponding hydroxylamine, is important because of its stability and ease of formation from the hydroxyamino state. When the nitroxyl function is attached to carbon atoms without α-hydrogen atoms, it is sufficiently stable to be isolated. When an α-hydrogen is present a disproportionation reaction occurs to yield a hydroxylamine and a nitrone (scheme 5):

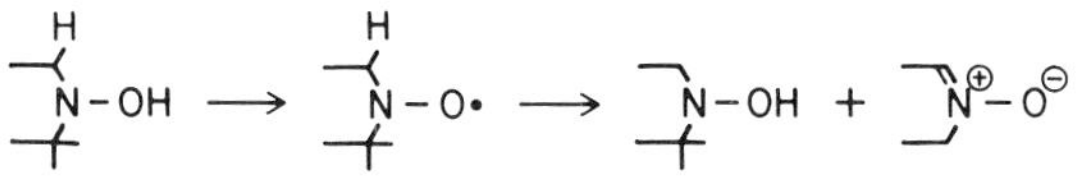

Nitroxides are readily formed from hydroxylamines and because of their stability are used as probes investigating free radical reactions. The formation of radicals in a number of biochemical reactions has been demonstrated by the conversion of a hydroxylamine to a stable nitroxide that could be detected by electron spin resonance spectroscopy. The hemolytic reactions of hydrazines [69], the generation of superoxide in microsomal incubations [60], one-electron oxidation pathways for cytochrome P450 [2], and the direct formation of nitroxides in microsomal incubations of carcinogenic aromatic amines [66] have been demonstrated with electron spin resonance spectroscopy utilizing the nitroxyl signal. In these studies, short-lived radicals reacted with the hydroxylamines to form the stable nitroxide which then accumulated and permitted detection.

Nitroxides are also formed by addition of a free radical to nitrones and when the nitrone is appropriately substituted the resulting nitroxide is sufficiently stable to be characterized. Nitroso compounds, such as nitrosobenzene and 2-methyl-2-nitrosopropane, react with free radicals to form nitroxides. The use of such reactions to stabilize radicals is called spin trapping [43].

Hydrazines

Chemical Properties

Alkyl- and aryl-hydrazines are generally weaker bases than the corresponding amines, ranging in pK_a from 5 to 8. Hydrazines are reactive nucleophiles, however, and are used to form derivatives of carbonyl functions called hydrazones. The 2,4-dinitrophenyl hydrazones of carbonyl compounds are classical derivatives for identification purposes and phenylhydrazine is used to convert sugars to osazones.

Oxidation

Hydrazines are also readily oxidized to diazenes in air and in the presence of mild oxidizing agents. Diazenes are also unstable and are readily oxidized to nitrogen and the alkane (scheme 6):

$$RNHNH_2 + [oxd] \longrightarrow RN{=}NH$$

$$RN{=}NH + [oxd] \longrightarrow RH + N_2$$

The second reaction is of toxicological interest because of the reaction of phenylhydrazine and hemoglobin. Hemoglobin will react with phenyldiazene formed from phenylhydrazine and oxygen, to generate nitrogen and an arylated heme. A radical reaction in which a complex, formed between the diazene and hemoglobin, decomposes to nitrogen, hydrogen atom and phenyl radical was proposed for the reaction [62] and has recently been demonstrated [17]. The two radicals can react with each other to form benzene or can react with another functional group to give hydrogen or phenyl adducts of other nearby functional groups. The formation of a phenyl-heme adduct and/or a reaction mediated by a radical generated in the heme-hydrazine reaction is thought to be the basis for Heinz body formation, a manifestation of phenylhydrazine toxicity. In support of this notion, Hill and Thornalley [36] have demonstrated that spin trapping agents which decrease the level of reactive radicals inhibit the hemolysis and hemoglobin oxidation caused by phenylhydrazine.

Nitroso Compounds

Chemical Properties

The nitroso state is a 2-electron oxidation of a hydroxylamine and is readily generated from hydroxylamines by air oxidation. A solution of an aromatic hydroxylamine will often contain the nitroso compound as a significant contaminant due to rapid air oxidation. This functional group is neutral and does not protonate, so that nitroso compounds have limited solubility in aqueous or polar solvents. The chemistry of the nitroso function in polar solvents is further complicated by reversible formation of a dimer [14] such that the concentration of the reactive species in any reaction can vary with the solvent [51]. The nitroso group is isoelectronic with dioxygen and binds to hemes in a similar manner. This property is the basis for the toxicity of aromatic nitroso compounds because they bind to hemoglobin [38] and form complexes with cytochrome P450 (see chapter III).

Aromatic nitroso compounds also react with olefins to form secondary hydroxylamines that are readily oxidized to nitroxides [67] (scheme 7):

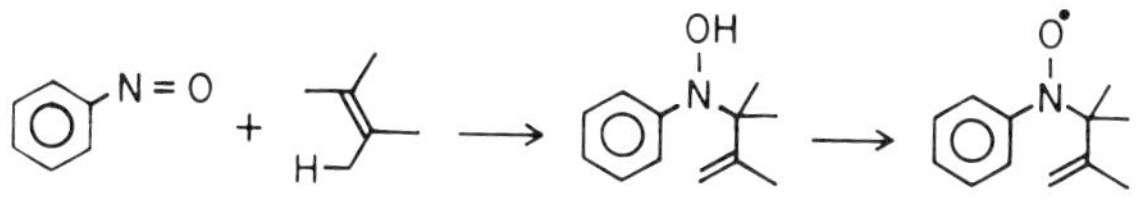

This pathway of radical formation has been proposed for the toxicity of nitroso compounds and their precursors [26]. According to this hypothesis the nitroso compounds add to the double bonds of lipids and form stable radicals on membrane surfaces.

Reduction

Aromatic nitroso compounds are reduced by NADPH to the hydroxylamines or azoxy compounds [6] the latter being condensation products of arylnitroso compounds and arylhydroxylamines. When nitrosobenzene is reduced with NADPH aerobically, the major product is the azoxy compound, whereas under anaerobic conditions the major product is phenylhydroxylamine. Nitroso compounds are also reduced with dithionite ($S_2O_6^{2-}$) to the hydroxylamines. Microsomal reduction of aliphatic C-nitroso compound to the corresponding hydroxylamines has been demonstrated [7].

Oximes

Hydroxylamine reacts with carbonyl groups to form oximes and with carboxyl groups to form hydroxamic acids. Oximes are of interest as metabolites of amines and are thought to form biochemically from amines by one of three different pathways: (a) N-oxidation of the imine [57]; (b) by rearrangement or tautomerism of a nitroso compound, or (c) by direct enzymatic oxidation of hydroxylamines (for discussion, see chapter VII).

Nitro Compounds

The nitro function is the highest oxidation state of organic nitrogen compounds. The α-hydrogen in nitroalkanes is acidic and in strongly alkaline solutions generates a carbanion. This property is reflected in the conversion of alkyl nitro compounds to the corresponding carbonyl derivative by the Nef reaction [52]. In this reaction the alkyl nitro derivative is converted to its conjugate base which is treated with strong acid (scheme 8):

$$R_2CHNO_2 + OH^- \longrightarrow R_2\overset{\ominus}{C}-NO_2$$

$$R_2C{=}NOOH \longrightarrow R_2C{=}O + N_2O + H_2O$$

Aromatic nitro groups have a strong electron withdrawing effect on other functional groups on the ring, so that amines are made weaker bases and acids made stronger with the substitution of nitro for hydrogen. Reduction is a common reaction of the nitro group chemically and biochemically. Aliphatic nitro compounds are readily reduced by metals, such as zinc to the hydroxyamino, or amine state. The products of aromatic nitro compound reduction depend on the pH of the reducing media. Under acid conditions nitroso, hydroxyamino and amino states can be obtained. At alkaline pH, reduction and coupling occur with azoxy and hydrazo benzene as the products. At pH 7.0 dithionite, the commonly used reagent to reduce cytochrome P450, reduces aromatic nitro compounds to nitroso, hydroxylamino and aniline derivatives.

References

1 Albert, A.; Serjeant, E.P.: Ionization constants of acids and bases, p. 140 (Wiley & Sons, New York, 1962).

2 Augusto, O.; Beilan, H.S.; Ortiz de Montellano, P.R.: The catalytic mechanism of cytochrome P-450. Spin-trapping evidence for one-electron substrate oxidation. J. biol. Chem. *257:* 11288–11295 (1982).

3 Bamberger, E.; Seligman, R.: Oxydation aliphatischer Amine vom Typus >CH:NH_2. Chem. Ber. *36:* 701–710 (1903).

4 Bamberger, E.: Sulfomonopersäure als Mittel zur Strukturbestimmung bei Aminen. Chem. Ber. *36:* 710–714 (1903).

5 Barnes, K.K.; Mann, C.K.: Electrochemical oxidation of primary aliphatic amines. J. org. Chem. *32:* 1474–1479 (1967).

6 Becker, A.R.; Sternson, L.A.: Nonenzymatic reduction of nitrosobenzene to phenylhydroxylamine by NAD(P)H. Bioorg. Chem. *9:* 305–312 (1980).

7 Bélanger, P.-M.; Grech-Bélanger, O.: Microsomal C-nitroso reductase activity. Biochem. biophys. Res. Commun. *83:* 321–326 (1978).

8 Brown, H.C.; Kanner, B.: Preparation and reactions of 2,6-ditert-butylpyridine and related hindered bases – a case of steric hindrance toward the proton. J. Am. chem. Soc. *88:* 986–992 (1966).

9 Burgen, A.S.V.: Comparative activity of arecoline and arecoline N-metho salt. J. Pharm. Pharmacol. *16:* 638 (1964).

10 Castellano, J.A.; Green, J.; Kauffman, J.M.: Aromatic polyfluoronitroso compounds. J. org. Chem. *31:* 821–824 (1966).

11 Cho, A.K.; Miwa, G.T.: Role of ionization in the N-demethylation of some N,N-dimethylamines. Drug Metab. Disposit. *2:* 477–483 (1974).

12 Clark, J.; Perrin, D.D.: Prediction of strengths of organic bases. Q. Rev. Chem. Soc. *18:* 295–320 (1964).

13 Cohen, G.; Collins, M.: Alkaloids from catecholamines in adrenal tissue: possible role in alcoholism. Science *167:* 1749–1751 (1970).

14 Coombes, R.G.: Nitro and nitroso compounds; in Sutherland, Comprehensive organic chemistry. The synthesis and reaction of organic compounds, vol. 2, pp 305–382 (Pergamon Press, New York 1979).

15 Dapo, R.F.; Mann, C.K.: Anodic oxidation of triethylamine. Analyt. Chem. *35:* 677–680 (1963).

16 Davis, V.E.; Walsh, M.J.: Alcohol, amines and alkaloids: a possible biochemical basis for alcohol addiction. Science *167:* 1005–1007 (1970).

17 Delaforge, M.; Battioni, P.; Mahy, J.P.; Mansuy, D.: In vivo formation of σ-methyl- and σ-phenyl-ferric complexes of hemoglobin and liver-cytochrome P450 upon treatment of rats with methyl- and phenyl hydrazine. Chem.-Biol. Interact. *60:* 101–114 (1986).

18 Dingell, J.V.; Sulser, F.; Gillette, J.R.: Species differences in the metabolism of imipramine and desmethylimipramine (DMI). J. Pharmac. exp. Ther. *143:* 14–22 (1964).

19 Dyke, S.F.: The chemistry of the enamines (Cambridge University Press, London, 1973).

20 Emmons, W.D.: The oxidation of amines with peracetic acid. J. Am. chem. Soc. *79:* 5528–5530 (1957).

21 Emmons, W.D.: Peroxytrifluoroacetic acid. I. The oxidation of nitroamines to nitramines. II. Anilines to nitrobenzenes. J. Am. chem. Soc. *76:* 3468–3470 (1954).

22 Fatiadi, A.J.: Active manganese dioxide oxidation in organic chemistry. Part II. Synthesis *3:* 133–167 (1976).

23 Ferris, J.P.; Gerwe, R.D.; Gapski, G.R.: Detoxification mechanisms. II. The iron-catalyzed dealkylation of trimethylamine oxide. J. Am. chem. Soc. *89:* 5270–5275 (1967).

24 Ferris, J.P.; Gerwe, R.D.; Gapski, G.R. Detoxification mechanisms. III. The scope and mechanism of the iron-catalyzed dealkylation of tertiary amine oxides. J. org. Chem. *33:* 3493–3498 (1968).

25 Feuer, H.: The chemistry of the nitro and nitroso groups (Interscience, New York 1969).

26 Floyd, R.A.; Soong, L.M.; Stuart, M.A.; Reigh, D.L.: Free radicals and carcinogenesis. Some properties of the nitroxyl free radicals produced by covalent binding of 2-nitrosofluorene to unsaturated lipids and membranes. Archs. Biochem. Biophys. *185:* 450–457 (1978).

27 Forrester, A.R.; Ogilvy, M.M.; Thomson, R.H.: Mode of action of carcinogenic amines. I. Oxidation of N-arylhydroxamic acids. J. chem. Soc. 1081–1083 (1970).

28 Fukuto, J.M. Di Stefano, E.W.; Burstyn, J.N.; Valentine, J.S.; Cho, A.K.: Mechanism of oxidation of N-hydroxyphentermine by superoxide. Biochemistry *24:* 4164–4167 (1985).

29 Geneste, P.; Kamenka, J.M.; Ung, S.N.; Herrmann, P.; Goudal, R.; Trouiller, G.: Conformational determination of phencyclidine derivatives in view of structure-activity correlation. Eur. J. med. Chem., Chim. ther. *14:* 301–308 (1979).

30 Gragerov, I.P.; Levit, A.F.: Mechanism of oxidation of aromatic amines and nitroso compounds by Caro's acid. Zh. Obshch. Khim. *30:* 3726–3731 (1960).

31 Hall, H.K., Jr.: Correlation of the base strengths of amines. J. Am. chem. Soc. *79:* 5441–5444 (1957).

32 Hallström, G.; Kammerer, R.C.; Nguyen, C.N.; Schmitz, D.A.; Di Stefano, E.W.; Cho, A.K.: Phencylidine metabolism in vitro: the formation of carbinolamine and its metabolites by rabbit liver preparation. Drug Metab. Disposit. *11:* 47–53 (1983).

33 Hammett, L.P.: Physical organic chemistry; 1st ed., pp. 184–228 (McGraw-Hill, New York 1940).

34 Hampson, N.A.; Lee, J.B.; Morley, J.R.; Scanlon, B.: Oxidations involving silver. V. The oxidation of primary aliphatic amines. Can. J. Chem. *47:* 3729–3736 (1969).

35 Hanin, I.; Jenden, D.J.; Cho, A.K.: Influences of pH on the muscarinic action of oxotremorine, arecoline, pilocarpine and their quaternary ammonium analogs. Mol. Pharmacol. *2:* 352–359 (1966).

36 Hill, H.A.O.; Thornalley, P.J.: The effect of spin traps on phenylhydrazine-induced haemolysis. Biochim. biophys. Acta *762:* 44–51 (1983).

37 Hine, J.: Structural effects on equilibria in organic chemistry, p. 183 (Wiley & Sons, New York 1975).

38 Hirota, K.; Itano, H.A.: Influence of ring substituents on the binding of nitrosobenzene by ferrohemoglobin. J. biol. Chem. *253:* 3477–3481 (1978).

39 Hoag, M.K.P.; Trevor, A.J.; Asscher, Y.; Weissman, J.; Castagnoli, N., Jr.: Metabolism-dependent inactivation of liver microsomal enzymes by phencyclidine. Drug Metab. Disposit. *12:* 371–375 (1984).

40 Horner, L.; Anders, B.: Studies on the course of substitution. XXII. The polar and radical course of the reaction of diacetyl peroxide with primary, secondary and tertiary amines and with phosphines, arsine, thioethers, and ethers. Chem. Ber. *95:* 2470–2484 (1962).

41 Ibne-Rasa, K.M.; Edwards, J.O.: The mechanism of the oxidation of some aromatic amines by peroxyacetic acid. J. Am. chem. Soc. *84:* 763–768 (1962).

42 Jaffé, H.H.: Reexamination of the Hammett equation. Chem. Rev. *53:* 191–261 (1953).

43 Janzen, E.G.: A critical review of spin trapping in biological systems; in Pryor, Free radicals in biology, vol. 4, pp. 115–154 (Academic Press, New York 1980).

44 Kamenka, J.M.; Geneste, P.: Synthesis, conformation, and physical properties of phencyclidine and its derivatives; in Domino, PCP (Phencyclidine): historical and current perspectives, pp. 47–82 (NPP Books, Ann Arbor 1981).

45 King, C.M.; Allaben, W.T.: Arylhydroxamic acid acyltransferase; in Jakoby, Enzymatic basis of detoxication, vol. 2, pp. 187–197 (Academic Press, New York 1980).

46 Koritskii, A.T.; Lukornikov, A.F.: Formation of diarylamine oxide radicals in reactions between amines and peroxide radicals. Dokl. Akad. Nauk SSSR *147:* 1126–1129 (1962).

47 Lindeke, B.: Nonenzymatic rearrangement and coupling reactions of N-oxidized compounds; in Gorrod, Damani, Biological oxidation of nitrogen in organic molecules. Chemistry, toxicology, and pharmacology, pp. 377–385 (Horwood, Chichester 1985).

48 Lindeke, B.; Cho, A.K.: N-Dealkylation and deamination; in Jakoby, Bend, Caldwell, Metabolic basis of detoxication. Metabolism of functional groups, pp. 105–126 (Academic Press, New York 1982).

49 Lindsay-Smith, J.R.; Norman, R.O.C.; Rowley, A.G.: Amino oxidation. VI. Metal-ion induced dehydrative cyclisation, reduction and reductive fragmentation of arylethyldimethylamine N-oxides. J. chem. Soc., Perkin I, pp. 228–234 (1972).

50 Lindsay-Smith, J.R.; Norman, R.O.C.; Rowley, A.G.: Amine oxidation. VIII. Evidence for intramolecular hydrogen-atom transfer in amine radical cations. J. chem. Soc., Perkin I, pp. 556–571 (1973).

51 Lundkvist, G.; Jonsson, U.; Lindeke, B.: Decomposition rates of some trans-2-nitroso-1-phenylethane dimers in solution. J. chem. Soc., Perkin I (in press).

52 March, J.: Advanced organic chemistry. Reactions, mechanisms, and structure; 3rd ed., p. 786 (Wiley & Sons, New York 1985).

53 McEwen, C.M.; Sasaki, G.; Lenz, W.R., Jr.: Human liver mitochondrial monoamine oxidase. I. Kinetic studies of model interactions. J. biol. Chem. *243:* 5217–5225 (1968).

54 Millar, I.T.; Springall, H.D.: A shorter Sidgwick's organic chemistry of nitrogen (Clarendon Press, Oxford 1969).

55 Nelson, R.F.: Anodic oxidation pathways of aliphatic and aromatic nitrogen functions; in Weinberg, Techniques of electrochemistry, vol. 5, pp. 535–792 (Wiley & Sons, New York 1974).

56 Osawa, Y.; Coon, M.J.: Mechanism-based inactivation of rabbit liver microsomal cytochrome P-450 isozyme 2 by phencyclidine. Fed. Proc. *45:* 319 (1986).

57 Parli, C.J.; Wang, N.; McMahon, R.E.: The enzymatic N-hydroxylation of an imine. A new cytochrome P-450-dependent reaction catalyzed by hepatic microsomal monooxygenases. J. biol. Chem. *246:* 6953–6955 (1971).

58 Perrin, D.D.: Prediction of pK_a values; in Yalkowsky, Sinkula, Valvani, Physical chemical properties of drugs, pp. 1–48 (Dekker, New York 1980).

59 Prileschajew, N.: Oxydation ungesättigter Verbindungen mittels organischer Superoxyde. Chem. Ber. *42:* 4811–4815 (1909).

60 Rosen, G.M.; Finkelstein, E.; Rauckman, E.J.: A method for the detection of superoxide in biological systems. Archs. Biochem. Biophys. *215:* 367–378 (1982).

61 Russell, C.D.: Reaction of triethylamine at platinum in acetonitrile solution; solvent background with perchlorate supporting electrolyte. Analyt. Chem. *35:* 1291–1292 (1963).

62 Saito, S.; Itano, H.A.: β-Meso-phenylbiliverdin IXα and N-phenylprotoporphyrin IX, products of the reaction of phenylhydrazine with oxyhemoproteins. Proc. natn. Acad. Sci. USA *78:* 5508–5512 (1981).

63 Schill, G.; Borg, K.O.; Modin, R.; Persson, B.A.: Ion-pair extraction methods; in Bridges, Chasseaud, Progress in drug metabolism, vol. 2, p. 219 (Wiley & Sons, New York 1977).

64 Smith, P.A.S.: The chemistry of open-chain organic nitrogen compounds (Benjamin, New York 1965).

65 Smith, P.J.; Mann, C.K.: Electrochemical dealkylation of aliphatic amines. J. org. Chem. *34:* 1821–1826 (1969).

66 Stier, A.; Clauss, R.; Lueke, A.; Reitz, I.: Redox cycle of stable mixed nitroxides formed from carcinogenic aromatic amines. Xenobiotica *10:* 661–673 (1980).

67 Sullivan, A.B.: Electron spin resonance studies of a stable arylnitroso-olefin adduct free radical. J. org. Chem. *31:* 2811–2817 (1966).

68 Taft, R.W., Jr.: Separation of polar, steric, and resonance effects in reactivity; in Newman, Steric effects in organic chemistry, pp. 556–675 (Wiley & Sons, New York 1956).

69 Thornalley, P.J.: The homolytic reactions of 1-acetyl-2-phenylhydrazine and hydrazine: a spin trapping study. Chem. Biol. Interact. *50:* 339–349 (1984).

70 Tichý, M.; Jonáš, J.; Sicher, J.: Stereochemical studies. XVII. Dissociation constants of cyclohexanecarboxylic acids and cyclohexylamines and conformational equilibria. Coll. Czech. Chem. Commun. *24:* 3434–3440 (1959).

71 Ward, D.; Kalir, A.; Trevor, A.; Adams, J.; Baillie, T.; Castagnoli, N., Jr.: Metabolic formation of iminium species: metabolism of phencylidine. J. med. Chem. *25:* 491–492 (1982).

72 Wawzonek, S.; McIntyre, T.W.: Electrolytic oxidation of aromatic amines. J. electrochem. Soc. *114:* 1025–1029 (1967).

73 White, R.W.; Emmons, W.D.: Chemistry of permaleic acid. Tetrahedron *17:* 31–34 (1962).

A.K. Cho, PhD, UCLA Medical Center, Department of Pharmacology,
23–272 Center for the Health Sciences, 10833 Le Conte Avenue,
Los Angeles, CA 90024–1735 (USA)

Prog. bas. clin. Pharmac., vol. 1, pp. 27–62 (Karger, Basel 1988)

II. Cytochrome P450 in Nitrogen Metabolism

G.T. Miwa, J.S. Walsh

Department of Animal Drug Metabolism, Merck Sharp & Dohme Research Laboratories, Rahway, N.J., USA

I. Introduction

A. Historical

Compounds undergoing nitrogen oxidation by biological systems are usually transformed into stable, polar metabolites that are readily eliminated from the body in the urine. Other compounds, however, undergo N-oxidation to form unstable products, as may be the case for some N-dealkylation reactions, or react with water or biological nucleophiles to give secondary products. Much of the research on N-oxidation, particularly aromatic N-oxidations, arises from the interest in those secondary products that are formed from reactions with DNA and proteins since much of the carcinogenicity and toxicity of N-containing compounds can be attributed to the alkylation of these critical cellular macromolecules by reactive metabolites. Thus, the earliest reports on the biological transformation of N-containing compounds were primarily spurred by the desire to determine the cause of the toxicity and carcinogenicity of aromatic amines. For example, Hustedt and Kiese [78] demonstrated the four-electron oxidation of aniline to nitrosobenzene in the dog. A year later, Cramer et al. [27] demonstrated the two-electron oxidation of the carcinogenic amide, 2-acetylaminofluorene to form N-hydroxyacetylaminofluorene.

Less than a quarter century after these pioneering studies a vast body of data has become available on N-oxidation and reduction reactions in biological systems. An excellent historical review of this subject has been written by Jenner [82] and more recently by Uehleke

[134] and by Coutts and Beckett [25]. A recent review article by Lindeke [91] discusses the chemistry of N-oxygenated compounds. The flavin-containing monooxygenase is also a ubiquitous enzyme and catalyzes the N-oxygenation of a variety of secondary and tertiary amines. The enzyme has been isolated from the liver and lung of various species and the chemical and kinetic mechanisms elucidated in great detail. The reader is directed to a recent review by Ziegler [150].

B. Criteria for Cytochrome P450 Involvement

The purpose of this chapter is to describe oxidation and reduction reactions catalyzed by cytochrome P450 (P450). This has been an awkward task as the majority of research on N-metabolism has not focused specifically on demonstrating P450 involvement in these reactions. Thus, the exclusion of studies that have not rigorously demonstrated P450 involvement would have made this chapter very short indeed. Consequently, we have been quite lax with our criteria for P450 involvement.

A review of the literature indicates that a variety of criteria have been applied in order to implicate P450 in a particular reaction. Frequently, the induction and inhibition of metabolism of a compound by known P450 inducers, such as phenobarbitol (PB), 3-methylcholanthrene (MC), aroclor 1254, and inhibitors – SKF 525A, metyrapone, carbon monoxide (CO), 2,4-dichloro-6-phenylphenoxyethylamine (DPEA), 7,8-benzoflavone and *n*-octylamine – are taken as evidence for the involvement of P450. Other criteria have also been used, such as the binding spectrum resulting from the substrate-induced changes in the P450 spectrum, the microsomal localization of enzyme activity, and NADPH and O_2 requirements. A more rigorous criterion, although less commonly employed, is the production of a photoactivation spectrum which demonstrates the light reversal of the CO inhibition with a maximum effect at about 450 nm [24]. Finally, the most direct and rigorous methods for implicating P450 in a metabolic process are the inhibition of the reaction by specific antibodies to P450 or the demonstration that the reaction is catalyzed by purified isozymes of P450.

We have tried to be consistent in requiring the demonstration of at least one of these criteria for the citations noted in this chapter. However, we have abandoned even this requirement, in some cases, in order to include some of the earliest studies on N-oxidation as these studies date back to the discovery of P450 itself [88].

II. The Cytochrome P450 System

The hepatic P450 system is composed of at least two protein components, NADPH-cytochrome P450 reductase (F_p) and P450. The flavoprotein reductase, F_p, contains both flavin adenine dinucleotide (FAD) and flavin mononucleotide (FMN) prosthetic groups and functions in the transfer of reducing equivalents from NADPH to the cytochrome during oxygen activation [138]. This protein (MW 79,000) has been purified from a variety of sources [33, 40, 74, 149] and occurs at a low mole ratio (approximately 1:25) relative to P450 in the microsomal membrane. Various studies have shown that F_p and P450 must interact by lateral diffusion in membrane systems to form a transient bimolecular complex in order for electron transfer to P450 to occur [103, 148].

The term P450 refers to a family of structurally related isozymes, all containing a protoporphyrin IX prosthetic group, that function in the activation of molecular oxygen which is required for substrate oxidation. The overall monooxygenase reaction catalyzed by this system is:

$$RH + NADPH + O_2 + H^+ \rightarrow ROH + NADP^+ + H_2O. \quad (1)$$

Figure 1 illustrates some features of the catalytic cycle of this system. Substrate, RH, binds to the oxidized cytochrome, I, to form the binary P450-substrate complex, II. This binding often perturbs the P450 spectrum in the visible region causing a characteristic substrate-binding spectrum. The P450-substrate complex undergoes a one-electron reduction catalyzed by the reduced F_p forming III. This reduced cytochrome can coordinate with either CO to produce a λ_{max} at 450 nm, unique to this hemeprotein, or to O_2 to form the ternary complex, IV, which absorbs in the visible region with a λ_{max} at 440 nm [143]. Dissociation of the superoxide anion can occur from this complex and, after dismutation, results in the formation of H_2O_2 [136].

A second electron from F_p or, in some cases, from cytochrome b_5 further reduces the ternary complex to what is formally a ferric peroxide complex, V. Cleavage of the diatomic oxygen bond generates a molecule of water and an 'active oxygen' form of P450, VI. The ability of P450 to facilitate this oxygen bond cleavage makes it unique among the cytochromes in generating an electron-deficient oxygen atom. The mechanism by which P450 is capable of effecting this oxygen activation is the subject of active research (section V). Similarly, little is known about the

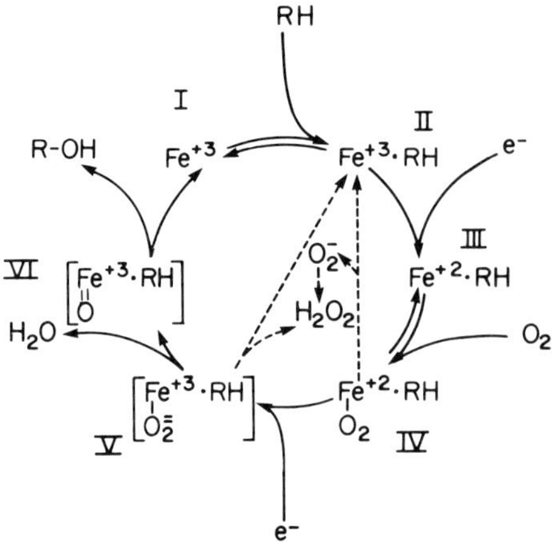

Fig. 1. Cytochrome P450 reaction cycle.

subsequent steps leading to the N-oxidized product although chemical model studies have provided some insight (section V) and accumulating recent evidence suggests that N-oxidation occurs by a step-wise one-electron transfer process (section VI).

III. N-Oxidation Reactions

Cytochrome P450 has been implicated in many N-oxidation reactions. Most primary amines can potentially undergo three successive two-electron oxidations to form the N-hydroxy, nitroso and nitro functionalities corresponding to oxidation states I, II and III in Figure 2[1]. Secondary and tertiary amines containing two and three nonlabile N-carbon bonds, respectively, can be successively oxidized to fewer products. For example, secondary amines can undergo a maximum of two, two-electron oxidations giving hydroxylamines or imines (oxi-

[1] In this chapter we employ a heuristic approach to categorizing the oxidation states of nitrogen rather than the functional categorization employed in chapter I. Accordingly, we have assigned the total oxidation state of the N- and α-carbon atoms to the N-atom. Thus, imines and oximes are described, respectively, as being formally 2- and 4-electron oxidation products of amine nitrogen (fig. 2) rather than the -3 and -1 valencies described in chapter I.

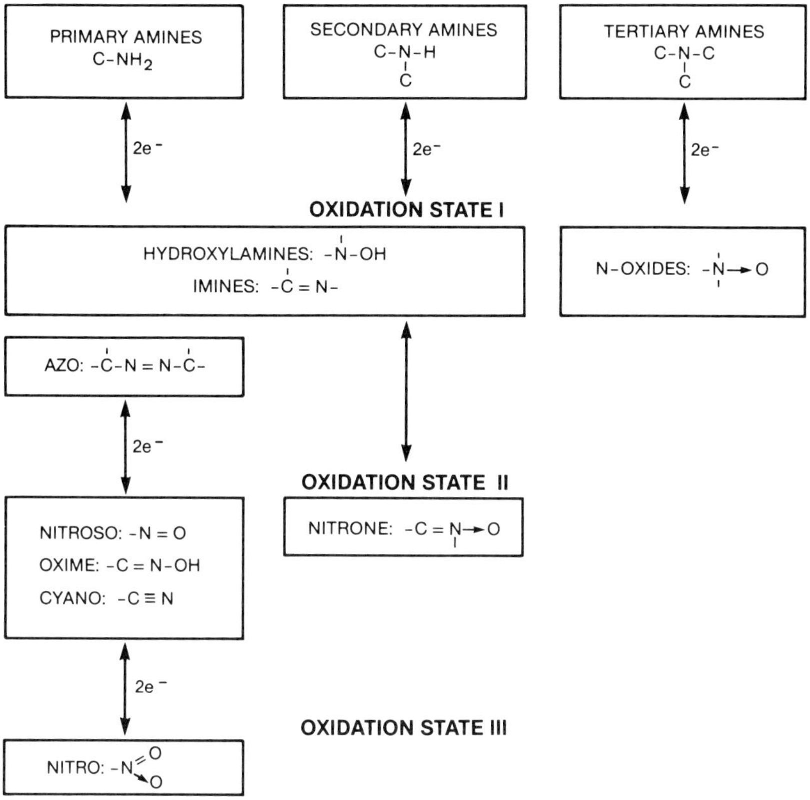

Fig. 2. Oxidation-reduction reactions of nitrogen-containing compounds.

dation state I) and nitrones (oxidation state II) while tertiary amines are generally able to undergo only a single, two-electron oxidation to form N-oxides (oxidation state I).

A. Oxidation State I

Aromatic and aliphatic primary and secondary amines and amides are N-hydroxylated by P450. Interest in this reaction has been stimulated by the toxicological and carcinogenic potential of hydroxylamines (see also chapters V, VI). Cramer et al. [27] were the first group to demonstrate N-hydroxylation in vivo. The aromatic amide 2-acetylaminofluorene, *1,* was N-hydroxylated, *2,* and subsequent studies by Johnson

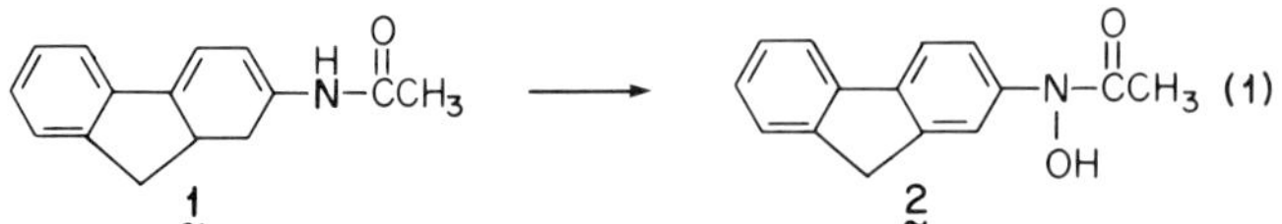

1 2

et al. [84] with purified enzymes have unequivocally demonstrated the involvement of P450 in this reaction.

The hydroxamic acid metabolite, *2,* was shown to be a proximate carcinogen or intermediary metabolite which, upon further biotransformation, resulted in the formation of the ultimate carcinogen, currently thought to be the sulfate ester of *2,* 2-acetylaminofluorene-*N*-sulfate [99]. Other examples are known for the activation of N-containing compounds by N-hydroxylation, such as the N-hydroxylation product of the antimalarial drug dapsone, *3,* which causes methemoglobinemia [28] and the activation of the hepatocarcinogen, N-methyl-4-aminoazobenzene *(4)* by N-hydroxylation of the desmethyl metabolite, *5* [85]. The hydroxylamine, *6,* is susceptible to nucleophilic attack and forms covalent bonds with proteins.

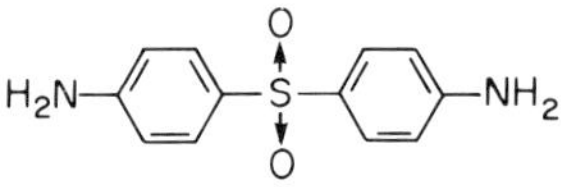

3

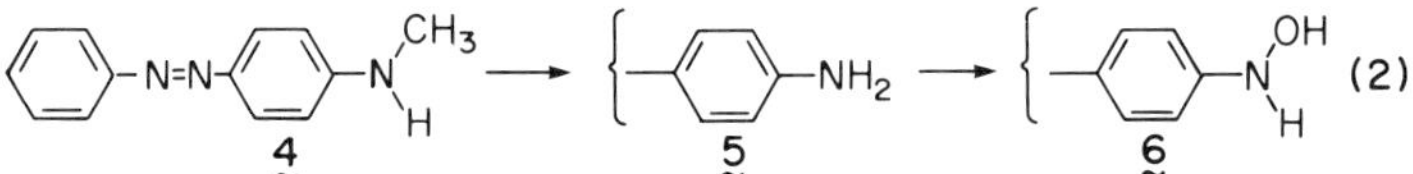

4 5 6

More recently, the aromatic amine hair dye component, 2,4-diaminoanisole, *7,* was shown to be metabolically activated by P450 to a reactive metabolite that covalently binds to proteins [37]. The binding is induced by PB and inhibited by CO. Antibodies directed against F_p also inhibited protein binding, implicating P450. When deuterium atoms were substituted for the methyl hydrogens, an inverse isotope effect was observed for protein binding, suggesting that the protein binding of *7*

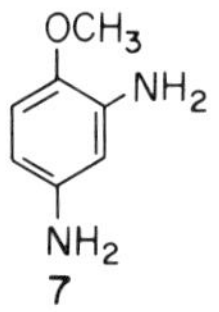

7

required metabolism at a site other than the 0-methyl group and that the deuterated analog had caused a change in regioselectivity away from the methoxyl group to this site. A metabolic pathway commencing with N-hydroxylation was consistent with these observations.

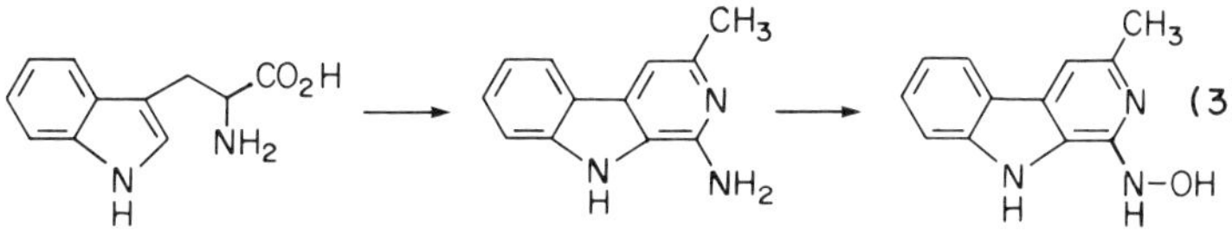

8 9 10 (3)

An excellent series of studies by Ishii et al. [79, 80] and Hashimoto et al. [63, 64] have shown that N-hydroxylation of various protein pyrolysates leads to the alkylation of DNA and mutagenicity. Thus, tryptophan, *8,* upon pyrolysis forms the indole derivative, 3-amino-l-methyl-^{5}H-pyrido [4,3-b]indole, *9* [63].

10 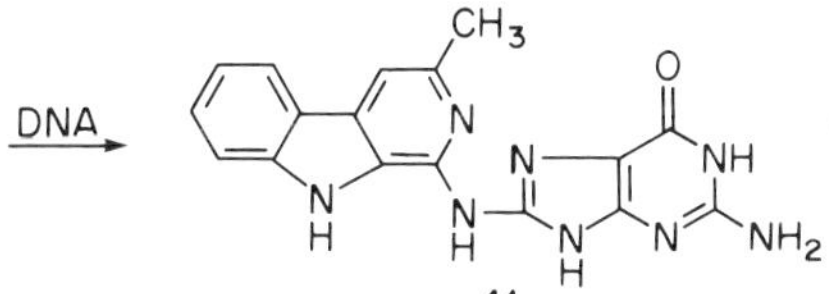

11 (4)

Mutagenicity resulting from N-oxidation catalyzed by P450 has been deduced from microsomal studies since this activity was induced by MC and was localized in microsomes. CO, 7,8-benzoflavone and N-octylamine inhibited the activation [79] while metabolically activated *9* alkylated DNA [64]. The N-hydroxylated metabolite, *10,* was identified and the structure of the guanine adduct of *10,* isolated from the alkylated DNA, was elucidated as 3-(8-guanyl)amino-l-methyl-5-pyrido[4,3-b]-indole, *11.*

12 13 (5)

A related study by these authors [80] has shown that the N-hydroxylated metabolite, *13,* of the glutamic acid pyrolysate product, 2-amino-6-methyldipyrido[1,2-a: 3′,2′-d]imidazole, *12,* is formed by purified isozymes of P450. Mutagenicity is dependent on P450, F_p and NADPH. The major metabolite, *13*, was identified and shown to be mutagenic without further metabolic activation.

Many other aromatic amines are activated to reactive intermediates that induce cellular damage either through alkylation of DNA or proteins. This subject is treated more extensively in chapters V and VI and in the review by Weisburger and Weisburger [142].

Although the availability of purified isozymes of P450 has greatly aided studies designed to implicate P450 in N-oxidation reactions, this subject has not always been so clear. Inhibitors such as metyrapone, which were used to implicate P450, gave conflicting results depending on the substrate examined. For example, metyrapone competitively inhibited *p*-chloroaniline and 2-naphthylamine N-hydroxylation [135] but did not inhibit N-methylaniline or *N,N*-dimethylaniline N-oxidation [134].

The controversy over the role of P450 in N-oxidation reactions also has its origins in the fact that microsomal preparations contain both P450 and the amine oxidase enzyme (chapter IV). Both enzymes require O_2 and NADPH and catalyze the N-oxidation of many tertiary amines to N-oxides. Based on literature observations and inhibition studies, Gorrod [48] postulated that substrates oxidized by these two enzymes could be distinguished based on their base strengths or pK_a values. Substrates with pK_a values below 1, such as amides, were proposed to be N-oxidized by P450 while more basic substrates with pK_a values greater than 7, such as aliphatic amines, were proposed to be N-oxidized by the amine oxidase enzyme. Substrates with intermediate basicity ($1<pK_a<7$), such as

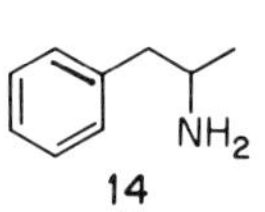

14

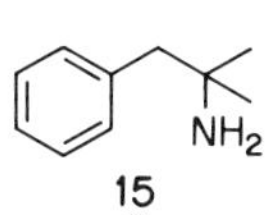

15

many aromatic amines, could be oxidized by both enzymes and, indeed, dimethylaniline has been shown to be N-oxidized both by purified preparations of the amine oxidase [151] and P450 enzymes [71, 72]. This hypothesis would then explain many of the apparent inconsistencies in elucidating the enzymes involved for numerous aromatic amines having intermediate pK_a values [48, 73, 134, 135]. However, the involvement of P450 in N-oxidation of amine substrates with pK_a values greater than 7 is still debated.

The research on amphetamine, *14,* and phentermine, *15,* metabolism illustrates this debate. The N-hydroxylation of both amines by liver microsomes was inhibited by CO and DPEA and was induced by PB [19, 20, 127].

In contrast, Beckett, Belanger and co-workers [4–6, 25] have concluded that this N-oxidation reaction was not catalyzed by P450 based on the stimulation by azide and cyanide which were considered to be inhibitors of the microsomal alcohol oxidase system but not of the amine oxidase system. These authors suggested that the flavoprotein amine oxidase enzyme was responsible for this reaction, consistent with Gorrod's hypothesis. In addition, these authors concluded that only C-oxidation was catalyzed by P450. Recently, however, Duncan et al. [35, 36] have demonstrated that purified P450 catalyzes the N-hydroxylation of the arylaliphatic amine, phentermine. Amphetamine and some of its congeners have also been shown to undergo N-hydroxylation in reconstituted systems containing P450 [Duncan and Cho, personal communication].

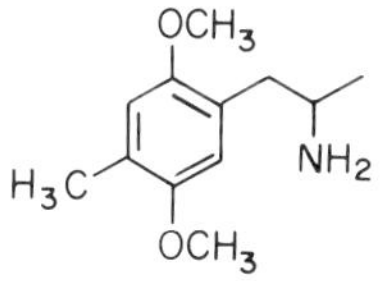

16

The *stereoselectivity* of the N-hydroxylation reaction of amphetamine [45] and the related compound, 1-(2,5-dimethoxy-4-methylphenyl)-2-aminopropane, *16* [44] have also been determined using pseudo-racemic mixtures composed of nondeuterated and deuterated enantiomers. Henderson et al. [69] examined the consequences of stereochemistry on the N-dealkylation and deamination of S(+) and R(−)-N-isopropylamphetamine The R(−) enantiomer had a 7-kcal/mol greater activation energy than the S(+), form suggesting that different mechanisms were involved. Moreover, a primary deuterium isotope effect was observed for the S(+) isomer when the α-hydrogen was substituted with deuterium while no isotope effect was observed for the R(−) isomer. The authors concluded that these reactions proceeded through the rate-limiting formation of a carbinolamine intermediate for the S(+) enantiomer and through N-oxidation for the R(−) isomer.

The two-electron oxidation of tertiary amines results in N-oxides (fig. 1, oxidation state 1). Historical reviews of N-oxide formation by

Table I. Examples of compounds forming N-oxide metabolites from liver microsomes

Name	Structure[a]	Reference(s)
Aliphatic amines		
Tamoxifen *(17)*	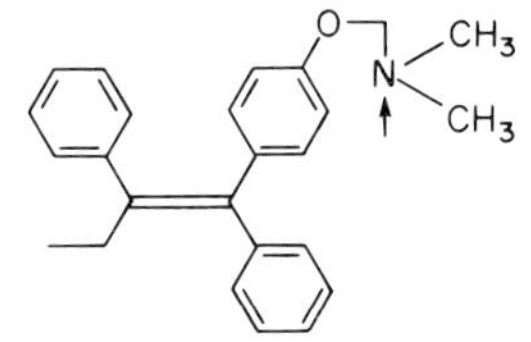	38
Imipramine *(18)*	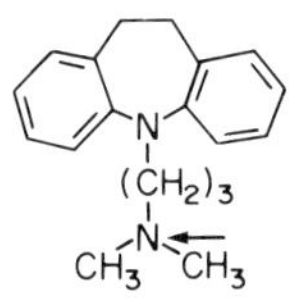	13
Dimethylaniline *(19)*	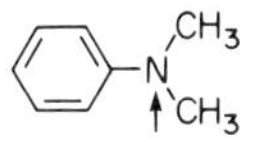	71–73, 146
N-*p*-Methoxyphenylaziride	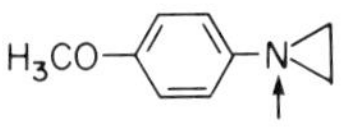	66
Aromatic amines		
Pyridine *(27)*		49
Nicotine *(28)*	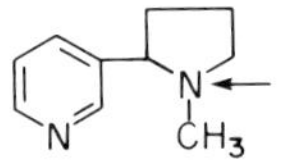	83
Cotinine *(29)*	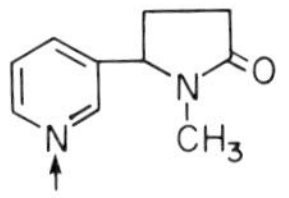	29

[a]Arrow indicates site of N-oxidation.

biological systems been written [12, 82] and the physical and pharmacological properties have been reviewed [12]. There is also a large literature demonstrating that microsomal enzymes produce N-oxides from either aliphatic or aromatic amines. For example, the N-oxides of the aliphatic amines, tamoxifen, *17* (table I), and imipramine, *18*, and the imipramine-related compounds amitryptyline [3], chloropromazine [13] and *N,N*-dimethyl-^{5}H-dibenzo-[a,d]cycloheptene-Δ^5,γ-propylamine [10] have been reported.

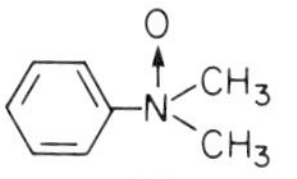

20

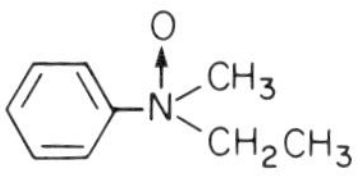

21

The N-oxides of aromatic amines, such as dimethylaniline-*N*-oxide, *20*, *N*-ethyl-*N*-methylaniline, *21* [50], and *N,N'*-bis(2-chloroethylene-*p*-anisidine), *22*, are also produced by microsomal enzymes (table I). The microsomal metabolism of N-mustards, such as *22*, results in the N-dealkylation product *23* and the aziridine, *24* [66]. The authors postulated an intermediate N oxide during the N-dealkylation reaction to form *23*, but a carbinolamine intermediate might be more plausible as other N-dealkylation reactions appear to proceed through this intermediate [147]. The aziridine, *24*, is metabolized further by microsomal enzymes to the nitroso compound, *26*, and, presumably, ethylene [66]. This latter reaction appears to proceed through the N-oxide intermediate, *25*, since aziridine N-oxides are known to decompose into nitroso and olefin products [65].

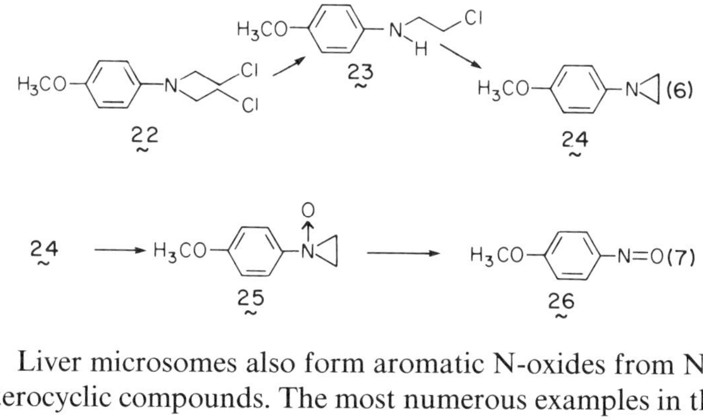

Liver microsomes also form aromatic N-oxides from N-containing heterocyclic compounds. The most numerous examples in the literature are N-oxides of pyridine derivatives. For example, pyridine, *27*,

3-methylpyridine [49] and 3-chloropyridine [66] form N-oxide metabolites. Similarly the nicotine derivatives cotinine, *29,* and N′-nitrosonornicotine [67] both form pyridine N-oxide metabolites in vitro, while nicotine, *28,* is N-oxidized on the pyrolidine ring to form nicotine-l′-N-oxide. The stereoselectivity of the latter reaction has been examined and shows that the N-oxide with the 1′-R configuration of the N-atom is preferentially formed [83]. Other examples of biologically produced N-oxides of aromatic heterocycles are described by Gorrod [47].

It was previously mentioned that the involvement of P450 in the N-hydroxylation of amine substrates is debated. Similarly, there are studies that both support [71, 72] and refute [3, 50, 73, 135] N-oxide formation catalyzed by P450. Studies of pyridine N-oxidation (*27* in table I) have implicated P450 in this microsomal reaction because N-oxide formation is inhibited by SKF 525A, DPEA and CO but is induced by PB [49]. A study by Hlavica and Hulsmann [71] with the cytochrome P448 isozyme purified from rabbit liver microsomes demonstrated that at least one isozyme of P450 was capable of catalyzing this oxidation. Thus, the more recent evidence indicates that P450 can catalyze N-oxide formation although there seems to be little disagreement that the flavin amine oxidase system contributes at least 50% to N-oxide formation in microsomal preparations [71, 72].

B. Oxidation State II

Metabolic oxidation of azo compounds such as dibenzo[c,f]-[1,2]-diazepine, *30* [41] and procarbazine [141] results in azoxy metabolites such as *31.* These are formally three electron oxidation products of nitrogen although the enzyme is actually only catalyzing the two electron oxidation of one of the azo N-atoms.

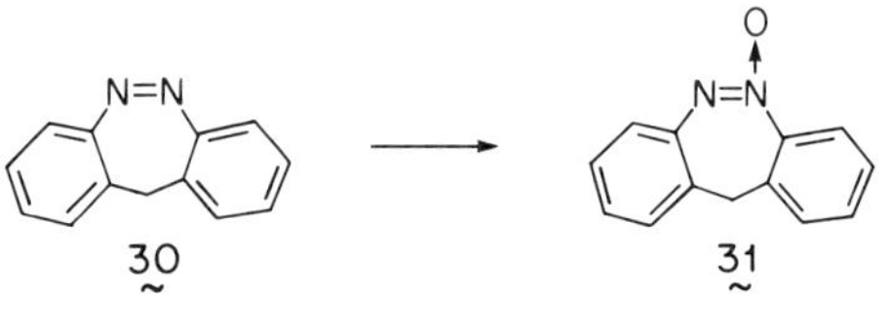

(8)

The formation of nitroso compounds from primary amines and nitrones from secondary amines are representative of the oxidation state II of nitrogen (fig. 1). 2-Methyl-2-nitroso-1-phenylpropane, *32,* is formed in microsomal incubations of phentermine, *15,* presumably by way of an intermediate hydroxylamine [4–6].

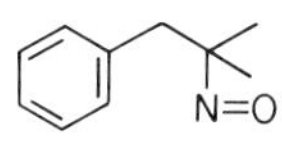

32

An active hydrogen α to the nitroso group can result in tautomerization to the oxime. For example, nitrosoamphetamine, *33,* readily tautomerizes to the oxime, *34,* which has been isolated from incubations of amphetamine with liver microsomes [92].

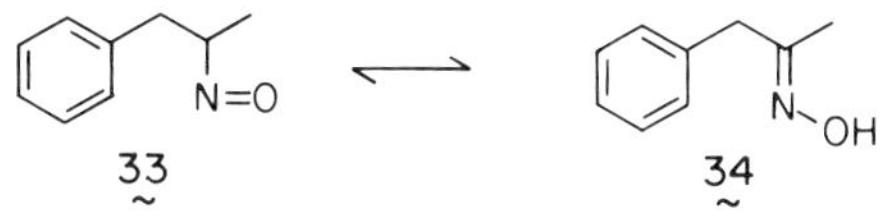

(9)

Nitroso compounds can also condense with themselves to form dimers. The dimers of 2-methyl-2-nitroso-1-phenylpropane, *32,* and 2-nitroso-1-phenylpropane, *33,* have been reported [8, 92].

Matsumoto and Cho [97] have demonstrated that N-hydroxyamphetamine, *33,* was oxidized to the oxime, *34,* by liver microsomes. This reaction was, however, not inhibited by CO, SKF 525A or DPEA, suggesting that P450 was not involved in this NADPH-dependent reaction.

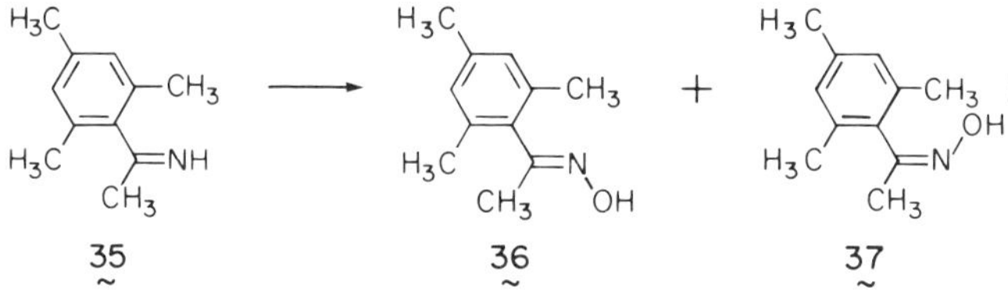

(10)

Another mechanism for oxime formation has been demonstrated by Parli et al. [111]. Imines can undergo P450-dependent oxidation to form oximes. The compound, 2,4,6-trimethylacetophenone imine, *35,* was metabolized to a 1:10 mixture of anti, *36,* and syn, *37,* oximes. Moreover, this microsomal reaction required NADPH and O_2, was induced by PB and was inhibited by CO, SKF 525A and DPEA. Tracer studies employing $^{18}O_2$ demonstrated that 97% of the oxime oxygen arose from molecular oxygen. This imine oxidation reaction, mediated by P450, supports an alternate mechanism for oxime formation. Thus, oximes could arise through the direct oxidation of an imine intermediate rather than through the stepwise oxidation of nitrogen to the nitroso compound followed by tautomerization to the oxime. Oxime formation from amphetamine has also been postulated to occur by oxidation of an imine intermediate [75, 76].

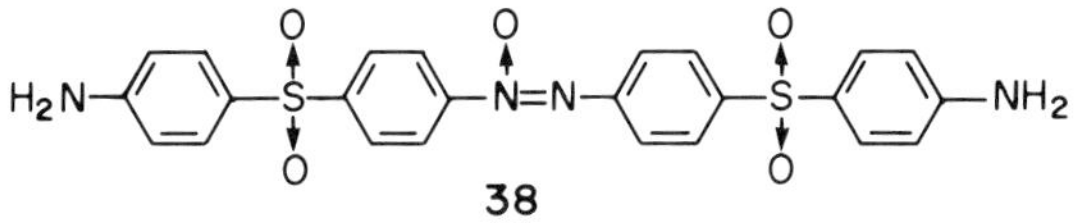

38

Azoxy compounds are, formally, three electron oxidation products of nitrogen and can arise through the two electron oxidation of an azo nitrogen, as discussed for dibenzo[c,f]-[1,2]diazepine, *30,* or through the condensation of a hydroxylamine (2 electron, oxidation state I) and nitroso compound (4 electron, oxidation state II). For example, the azoxy dapsone metabolite, *38,* was isolated from microsomal incubations of dapsone, *3* [28]. The hydroxylamine of dapsone was also identified and, although no nitroso dapsone was found, the authors suggested that the azoxy metabolite was formed through condensation of these two metabolites.

Similarly, the azoxy metabolite of *p*-thioanisidine has been observed. Oxygen label studies demonstrated that the azoxy oxygen originated from molecular oxygen and that CO inhibited azoxy metabolite formation, indicating that this was a monooxygenase reaction catalyzed by P450.

C. Oxidation State III

There are few examples of the biological oxidation of nitrogen to nitro compounds. The two electron oxidation of 1-phenyl-2-propanone oxime, *34,* to 2-nitro-1-phenylpropane, *40,* by liver 12,000 *g* supernatant was reported by Coutts et al. [26]. Beckett and Belanger [4, 5] have, however, suggested that the nitro compound, *40,* resulted as an artifact from the oxidation of the nitroso amphetamine, *33,* under the basic extraction conditions commonly used to isolate metabolites. This would not appear to be the case as both Coutts et al. [26] and Sum and Cho [128] have extracted and identified *40* under neutral pH conditions. Later studies by Beckett and Jones [9] have also verified this conclusion.

Sum and Cho [128] have also demonstrated the four-electron oxidation of *N-hydroxyamphetamine, 39,* to form *40* by liver microsomes.

Induction and inhibition studies suggested the involvement of P450. This group has also implicated the P450-dependent superoxide anion in

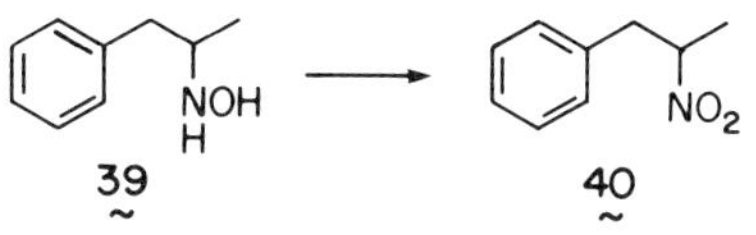
(11)

this reaction [98] since reaction 11 was inhibited by superoxide dismutase and a reconstituted purified P450 system effects this transformation [36]. The involvement of HO·, which could arise from O_2 by the Haber-Weiss reaction [59] or from the reaction of superoxide anion with reduced flavins [107] was excluded by studies with hydroxyl radical scavengers [98].

IV. Nitrogen Reduction

A. Nitro Reduction

Although many nitro compounds undergo reduction in the presence of microsomes [123, 131, 140] only a few studies have rigorously demonstrated P450 involvement [46, 62]. Cytochrome P450 participates in the reduction of *p*-nitrobenzoic acid and nitrobenzene to their corresponding amines by way of nitroso and hydroxylamine intermediates. Studies with purified enzymes have shown that the rate of reduction of nitrobenzene to aniline is limited by either the first or second two-electron reduction step. Consequently, formation of the nitroso or hydroxylamine intermediates appeared to be the rate-limiting step in the overall reduction to aniline. Moreover, the reduction of nitrobenzene exhibited a lag time which was not observed with nitrosobenzene or phenylhydroxylamine suggesting that the nitroso and hydroxylamine compounds were intermediates in this reaction.

Microsomal studies with antibodies to P450 and F_p and studies with the purified enzymes have shown that P450 is at least required for the final two electron reduction of phenylhydroxylamine to aniline. The reductase alone, however, was capable of catalyzing the first two, two-electron reduction steps to the nitroso and hydroxylamine intermediates. The role of P450 in these steps could not be determined. A relevant study was reported recently by Mansuy and Fontecave [96] using a chemical model system for P450 consisting of tetraphenylporphyrin and sodium ascorbate which catalyzed the reduction of both nitrobenzene and nitrosobenzene to aniline, suggesting that P450 may actually be capable of catalyzing the first two, two-electron reduction steps of nitrobenzene.

B. Azo Reduction

Azo reduction can be either flavoprotein or P450-mediated [70]. The flavoprotein, F_p, and P450 both catalyze azo reduction reactions although not all azo reductase activity can be attributed to these enzymes alone. For example, the azo reduction of neoprontosil, *41,* was inhibited 30–40% by CO and induced by either PB or MC, suggesting that P450 was at least partially responsible for azo reduction. However, not all of the MC-inducible activity could be abolished with CO and the increased levels of P450 did not directly correlate with azo reductase activity. Since MC does not induce F_p activity, these authors concluded that the increased azo reductase activity resulted from the induction of a third enzyme that was MC-inducible but insensitive to CO.

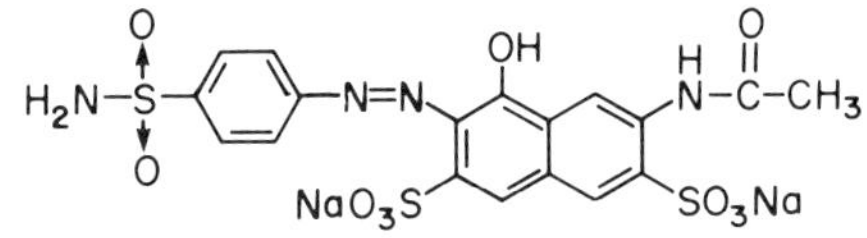

41

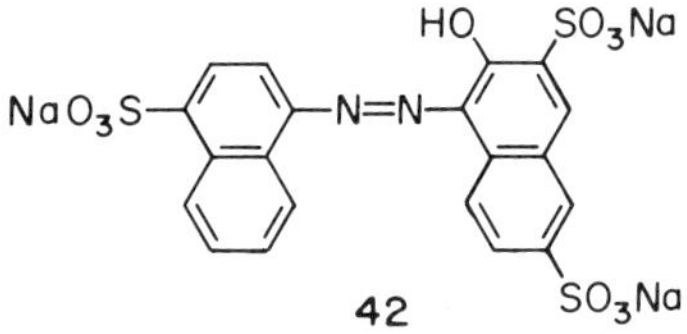

42

An opposite conclusion was reached by Fujita and Peisach [43]. When amaranth, *42,* was used as a substrate they observed a linear correlation between total P450 levels and azo reductase activity. Azo reductase activity was dependent only on the total P450 content in microsomal preparations rather than on the method of induction, suggesting that the P448 isozyme induced by MC was as efficient as the P450 isozyme induced by PB. A plot of azo reductase activity and P450 levels passed through the origin, suggesting that azo reduction was totally dependent on P450. Moreover, nonresponsive DBA/2N mice, which are genetically incapable of P448 induction by MC, also failed to increase azo reductase activity following MC treatment, suggesting that there was no MC-inducible non-P450 azo reductase.

The conflicting observations by different workers may result from the use of different substrates. Conceivably, amaranth may only be reduced by P450, whereas neoprontasil may be a substrate for a wider range of azo reductases. More recent studies by Fujita et al. [42] have confirmed the involvement of P450 and P448 in amaranth azo reduction through the use of specific antibodies. About 90% of the azo reductase activity was inhibited in PB- and MC-induced microsomes, respectively, when the P450 and P448 antibodies were included in the incubation mixture.

C. N-Oxide Reduction

There are a number of enzymes, in addition to P450, that are capable of N-oxide reduction. Among them are xanthine oxidase [106], enzymes in the intestinal microflora and heptatic cytoplasm [7]. In the rat, N-oxide reductase activity is associated primarily with the liver. The intestine has about 40% of the activity of the liver but contains negligible quantities of P450, suggesting the presence of an additional reductase in this tissue. This reductase activity was partially inhibited by propyl gallate, an inhibitor of F_p [133].

Studies with purified P450 have shown that both P450 and P448 will reduce tiaramide N-oxide, *43,* at comparable rates and that P450R isolated from rat, rabbit and *Pseudomonas putida* have comparable reducing activity despite widely different substrate specificities in oxidation reactions [81]. This was interpreted to indicate that the N-oxide did not bind to the normal substrate-binding site but interacts directly with the cytochrome heme itself. Reduced methyl viologen can replace NADPH in the N-oxide reduction catalyzed by microsomal systems [81, 86, 126]. Moreover, studies with purified enzymes have demonstrated that methyl viologen could also replace F_p and could accelerate N-oxide reduction by 7000-fold [81]. Consequently, the reduced methyl viologen appears to directly cause the reduction of the P450 heme which is necessary for catalytic N-oxide reduction.

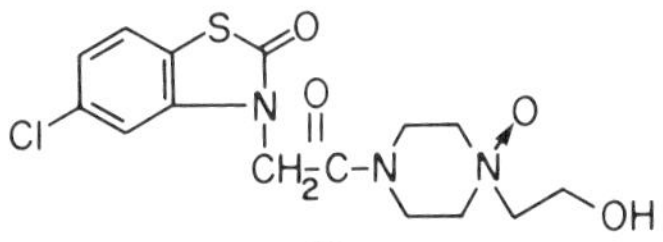

43

The biochemical pathway for N-oxide reduction was further elucidated by Sugiura et al. [126]. When tiaramide N-oxide, *43,* imipramine N-oxide, *18,* and dimethylaniline N-oxide, *21,* were used as substrates, microsomal reduction to the tertiary amines was inhibited by CO. Removal of NADH-cytochrome b_5 reductase or cytochrome b_5 from microsomes with trypsin or subtilisin reduced the NADH-dependent activity by 77%. Imipramine N-oxide reduction was inhibited 85% by antibodies against cytochrome b_5 reductase. These observations clearly implicated NADH-cytochrome b_5 reductase and cytochrome b_5 in the electron transport process leading to N-oxide reduction.

It appears that there are multiple forms of P450 with variable N-oxide-reducing capabilities. This is based on the observation that only some inhibitors of P450 will inhibit N-oxide reduction [30, 126] and that N-oxide reductase activity cannot always be correlated with total P450 levels after MC or PB induction [114, 126].

A unique NADH-dependent P450 isozyme has been proposed for indicine N-oxide reduction [114]. Low concentrations of KCN inhibited NADH-dependent but not NADPH-dependent reduction. Inhibition of the NADH supported reduction occurred at KCN concentrations below that required to inhibit oxidase activity. Acetone stimulated NADPH-dependent reduction but inhibited NADH-dependent reduction. Although both NADPH- and NADH-supported reactions gave similar K_m and V_m values, PB induction produced a selective increase in V_m for the NADPH-dependent reduction, while no change was observed in V_m or K_m for the NADH-supported reaction. These observations were rationalized in terms of a unique isozyme of P450 which could accept electrons from NADH to effect the N-oxide reduction but which would not contribute to oxidative drug metabolism.

The mechanism of N-oxide reduction is thought to require direct coordination of the N-oxide to the iron of P450. This was based on the observation that N-oxide reduction was inhibited by ligands capable of direct coordination with the heme iron such as CO, O_2, and type II ligands, such as aniline and octylamine. Moreover, tiaramide N-oxide, *43,* produced spectral changes with reduced P450 which caused a new peak of 442 nm [87, 125], and imipramine N-oxide reduction could be effected with only ethylenediaminetetraacetic acid and ferrous iron. The sequence of events is currently thought to involve (a) reduction of P450 by NADPH via F_p, (b) direct coordination of the N-oxide substrate to the heme of P450, and (c) a further one-electron reduction of the complex followed by the two-electron reduction of the N-oxide to the tertiary amine and water [125].

V. Chemical Models for Cytochrome P450-Catalyzed Oxidation/Reduction Reactions

The mechanism of P450-catalyzed N-oxidation reactions is only partially understood. Although a number of N-oxidation intermediates have been isolated and pathways interconnecting these intermediates proposed, the nature of the actual P450-oxidizing species is the subject of ongoing investigations.

One approach to mechanistic investigations is through chemical model systems where the nature of the oxidizing species is more clearly defined. A full review of model P450 systems is beyond the scope of this chapter and the reader is referred to recent review articles on this subject [34, 51, 56, 57].

$$Fe^{3+} \xrightarrow{e^-} Fe^{2+} \xrightarrow{O_2} \underset{\mathbf{44}}{[Fe^{3+}O_2^-]} \xrightarrow{e^-} \underset{\mathbf{45}}{[Fe^{3+}O_2^{2-}]} \quad (12)$$

It is now established that P450 combines with molecular oxygen after reduction to the ferrous state (equation 12). Mössbauer studies [119] on the bacterial P450-oxygen complex, however, suggest that the iron is in the ferric oxidation state at this stage. Thus, a ferric-superoxide complex, *44,* would be a more accurate representation of this complex. This is reduced further by one electron to give a ferric peroxide complex, *45.*

The precise structure of the ferric peroxide complex, *45,* and its involvement in substrate oxidation are less well established.

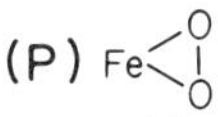

46

Poryphyrin peroxide complexes analogous to that proposed for P450 have been prepared. Electrochemical reduction of ferrous octaethylporphyrin in *t*-butylammonium perchlorate buffers in the presence of molecular oxygen produced a species spectroscopically consistent with an iron porphyrinperoxide complex. EPR and IR indicate this to be a high-spin η^2 ferric peroxide complex, *46.* Such complexes, where oxygen acts as a formal four-electron donor, are well known for a number of transition metals [22].

The optical spectrum, however, suggested a six-coordinate ferrous complex with a strong π-acceptor η^2-oxygen ligand. The back bonding of metal *d*-orbital electrons into the antibonding orbitals of the dioxygen ligand would weaken the oxygen-oxygen bond and facilitate bond cleavage. The oxygen bond cleavage step could occur either heterolytically or homolytically [89]. A homolytic radical mechanism for oxygen bond cleavage has been proposed for P450-mediated hydroxylation reactions based on product formation from a peroxyacid model for O_2 [45]. Isotope effect measurements are also consistent with this proposal [55]. Other studies with O_2-porphyrin complexes and chemical reducing agents suggest it is O_2-porphyrin reduction that is the rate-limiting step in the overall process of oxygen activation and substrate oxidation [129].

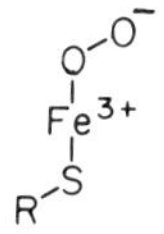

47

Magnetic circular dichroism studies with bacterial P450 and iron protoporphyrin IX dimethyl ester ligated to aromatic thiolates suggested that a sulfur was ligated to iron when substrate binds to P450 [31]. The authors further proposed that this axial ligand was retained when oxygen bound and, upon reduction, gave the $\eta^1Fe^{3+} \cdot O_2^{2-}$ complex, *47.*

Although transition metal-dioxygen complexes are capable of oxidizing substrates within the coordination sphere of the metal [137], it is generally believed that for P450-peroxide complex, addition of two protons results in the loss of water to form the ferric oxene, *48* [17, 31], which then participates in the oxidation of the substrate.

$$(P)Fe^{3+}:\ddot{O} \longleftrightarrow (P)Fe^{4+}:\dot{O}\cdot \longleftrightarrow (P)^{(+)}Fe^{4+}:\ddot{O}: \quad (13)$$

48 49 50

The oxene is formally equivalent to compound I of catalase and peroxidase [105]. Such a species, *48,* arising from Fe^{3+}, has the iron in a formal oxidation state of V with an eight-electron oxygen ligand, as $Fe^{5+} \cdot O^{2-}$. The actual electronic distribution, however, is unknown an Fe^{3+} six-electron oxygen complex, Fe^{4+} seven-electron oxygen complex, or an Fe^{4+} eight-electron oxygen complex with the porphyrin or fifth ligand sulfhydryl as a radical cation (equation 13) can be envisioned [17]. These structures differ only in the electron distribution within the complex. None of these species has been observed directly because of the instability of the porphyrin-oxygen complex [21].

Analogous species have been characterized, however, for other transition metals. For example, a stable oxoporphinato chromium V complex has been prepared [52] and is stable at ambient temperature for several hours. The infrared spectrum showed a Cr=O stretching frequency at 1,026 cm^{-1}. Subsequent work has demonstrated the ability of oxochromium porphyrins to oxidize organic substrates [53]. EPR studies with a system composed of tetraphenylporphinato manganese (III), and iodosyl benzene, and this system's reaction with alkanes, have suggested the formation of a hydroxymanganese (IV) porphyrin complex. This complex is thought to result from hydrogen abstraction from the substrate by the oxo intermediate. Hydroxylation of the alkane could then be envisioned to occur by hydroxyl radical ligand transfer to the alkane radical [122].

Alternative mechanisms to the peroxide-oxene mechanism have been proposed. Studies with bacterial $P450_{cam}$ suggested the participation of a free carboxylic acid in the catalytic cycle which led to the formation of an enzyme-bound peracid [60, 121]. The peracid could then oxidize the substrate directly or decompose to the ferric oxene intermediate [121]. In the bacterial $P450_{cam}$ system this carboxylic acid was shown to be a glutamic acid residue in the protein putidaredoxin which exists as a complex with P450. Other carboxylic acids can replace the putidaredoxin glutamate residue in catalytic activity. In addition, studies using lipoic acid and $^{18}O_2$ suggested a mechanism involving the direct oxidation of the substrate camphor by the enzyme-bound peracid. Other oxygen activation mechanisms are discussed in the review by White and Coon [144].

Mammalian P450 is believed to have a cysteine thiol as an axial ligand [31]. Recent X-ray crystallographic studies have definitely identified this ligand as cys 357 in bacterial $P450_{cam}$ [112]. This sulfur ligand is unique for cytochromes and is thought to play an active role in the activation of molecular oxygen. Several model P450 systems containing heme axially ligated to sulfur have been prepared [11, 15, 16, 23, 31]. These model systems all demonstrated the unique CO-reduced spectrum of P450 with a λ_{max} near 450 nm. For example, protohemin treated with mercaptans and base gave, under a CO atmosphere, a Soret band with an λ_{max} at 450 nm [15]. Treatment of protophoryrin IX diethyl ester with sodium methylsulfide in the presence of a crown ether gave a CO complex with an λ_{max} at 449 nm and an extinction coefficient similar to P450. When propyl mercaptan was used in place of sodium methylsulfide, a CO complex at 422 nm was obtained, suggesting that the 450-nm Soret band of the CO complex was due to a mercaptide rather than a mercaptan ligand [23].

Model studies with protohemin and *n*-butyl mercaptide suggested that the dioxygen complex of P450 did not contain a mercaptide ligand

[16]. Although this complex tended to undergo autooxidation, at −45 °C, two Soret peaks were observed at 476 and 378 nm which were consistent with O_2 binding. In contrast, the P450-oxygen complex showed a single peak at 418 nm. Furthermore, unlike other heme molecules [21], it was found that the mercaptide-heme complexes had comparable affinities for O_2 and CO. Chang and Traylor [18] have suggested that the interaction of sulfur *p*-orbitals with the iron resulted in π-electron donation from sulfur to iron and subsequently to the π*-orbitals of dioxygen. The differences in optical spectra of the model compound and P450 were, therefore, taken as evidence for the absence of a mercaptide axial ligand when O_2 binds to P450. The authors postulate that either protonation of the mercaptan ligand occurs or that replacement of the mercaptide with a nonsulfur ligand occurs when dioxygen binds to reduced P450. More recently, model porphyrins containing trans thiolate and oxygen ligands have been shown to mimic the low-spin ferric P450, as characterized by distinct α- and β-bands at 530 and 560 nm in the UV spectrum [116].

Other model systems have concentrated on mimicking the reactions catalyzed by P450 [22, 58]. Systems comprised of chlorotetraphenylporphinato iron ([TPP]FeCl) [56, 57] or chlorodimethyl ferriprotoporphyrin with iodosobenzene as an oxygen source catalyzed the epoxidation of olefins and hydroxylation of hydrocarbons [58]. A 'strapped' porphyrin with a hydrocarbon bridge across one face of the porphyrin ring, while the other face was masked as a μ-oxo dimer to prevent autooxidation, catalyzed the intramolecular oxidation of this hydrocarbon bridge in the presence of iodosoxylene [16]. Takeuchi et al. [130] have recently synthesized other model systems capable of binding small substrate-like molecules by hydrophobic host-guest interactions.

Unlike carbon oxidation models there are few examples of P450 models that mimic its nitrogen oxidation/reduction reactions. Santa et al. [117] effected oxidative N-dealkylation of anilines and oxidation of sulfides, using a system comprised of (TPP)FeCl, O_2 and $Na_2S_2O_4$ as reducing equivalent. In this case, porphyrin μ-oxo dimer formation, often a major source of loss of catalyst, was suppressed by the use of a mixed homogeneous protic solvent system. Tetraphenylporphinato iron (II) has been used to effect reduction of a number of aliphatic and heterocyclic N-oxides to the corresponding free amines. The resulting oxotetraphenylporphinato iron IV could then be reduced by triphenylphosphine to produce a catalytic cycle [104].

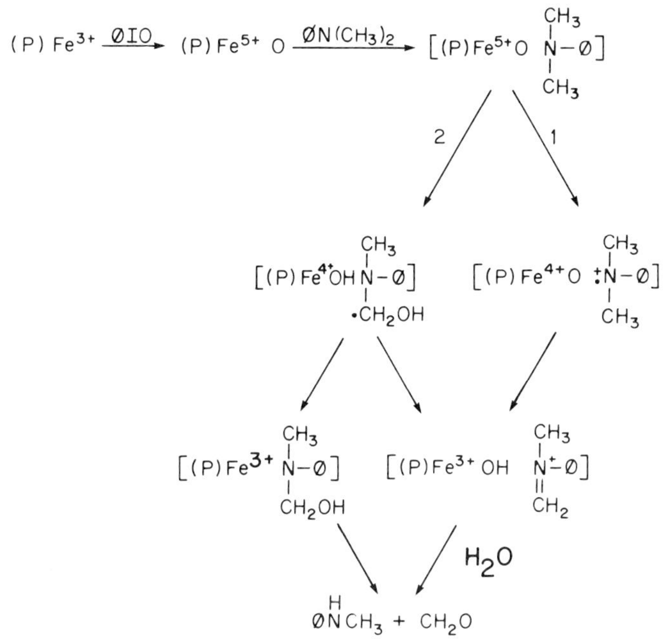

Fig. 3. Possible mechanisms for N-dealkylation reactions catalyzed by Heme model systems.

Heme model systems that catalyze oxidative N-dealkylations have also been developed by Bruice and co-workers [32, 118]. Reactions of tetraphenylporphinato iron with dimethylaniline and iodosobenzene resulted in the formation of N-methylaniline and formaldehyde [118]. Alternatively, dimethylaniline and iodosobenzene could be replaced with dimethylaniline N-oxide to yield the same products. As with other iodosobenzene-supported model reactions [58], an intermediate, $(P)Fe^{5+} \cdot O^{2-}$, is proposed to form. The authors proposed that this intermediate could potentially react with dimethylaniline by two possible routes: (1) One-electron oxidation of nitrogen by the ferric oxene to yield an aniline cation radical (fig. 3, path 1) followed by a hydrogen atom transfer from N-methyl carbon to oxygen to form an iminium ion. Hydrolysis of this intermediate would produce the N-methylaniline and formaldehyde products. (2) Direct transfer of a hydrogen atom to the ferric oxene to form a carbon radical intermediate which can then undergo a second one-electron oxidation to form the iminium ion or combine with the nascent iron-bound hydroxyl radical to give a carbinolamine has been proposed by Groves and McClusky [54] for aliphatic carbon oxidation reactions catalyzed by P450. Both intermediates would then give rise to the observed products. Using *p*-cyanodimethylaniline-*N*-oxide (*p*-CDMAO), as both oxygen source and substrate, both iron and manganese porphyrin catalysts effect oxidative demethylation in benzonitrile or methylene chloride [32, 113]. With the maganese porphyrin in benzonitrile, N-demethylation to *p*-cyanomonomethylaniline occurs directly from the CDMAO·Mn(III)·TPPX complex. The rate of product formation was found to be highly dependent on the nature of the axial ligand. However, the iron catalyst, (TPP)Fe(III)Cl, produced *N*-formyl-*p*-cyano-*N*-methylaniline and *p*-cyano aniline products, suggesting the involvement of a free [(TPP)Fe(IV) = O] species in their formation.

A large number of model systems have been developed which mimic P450 in either its spectroscopic or metabolic properties. Few, however, have concentrated on the N-oxidation/reduction capabilities of P450. Such studies would prove useful in the elucidation of the mechanism of P450-mediated N-oxidation/reduction reactions.

VI. One-Electron Nitrogen Oxidations

Recent indirect evidence has been obtained for the one-electron oxidation of nitrogen catalyzed by P450. The radical cation, *52,* resulting from such an oxidation best rationalizes the chemistry required to explain a variety of products.

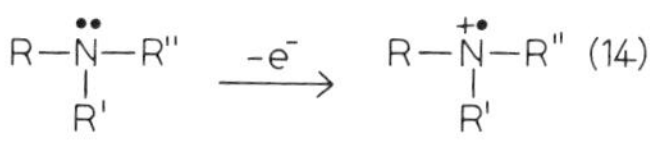

51 52

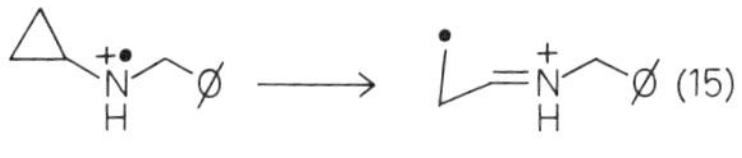

53 54

Thus, the homolytic opening of the cyclopropyl ring of an intermediate cation radical, *53,* to yield the carbon radical, *54,* has been implicated in the suicidal inactivation of P450 by benzylcyclopropyamine [61, 95]. The negative slope of the Hammett plot of the N-demethylation of a series of para-substituted *N,N*-dimethylanilines also indicates that this reaction is facilitated by electron donating substituents [14].

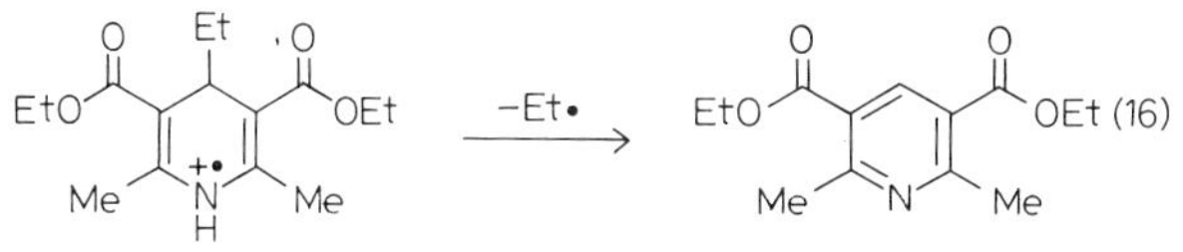

55 56

Alkyl radical formation from 4-alkyldihydropyridines has been documented and provides the most direct evidence for a cation radical, *55*, formed by one-electron oxidation of nitrogen [2].

Evidence from other elimination reactions also support initial one-electron N-oxidations catalyzed by P450. For example, an intermediate imine methide, *58*, resulting from hydrogen elimination from the cation radical, *57*, is suggested by the observed glutathione adduct, *59*, [109].

Similarly, proton-elimination from the cation radical of *p*-cyano-*N*-methylaniline rationalizes the N-methyl glutathione conjugate observed by Logan et al. [94].

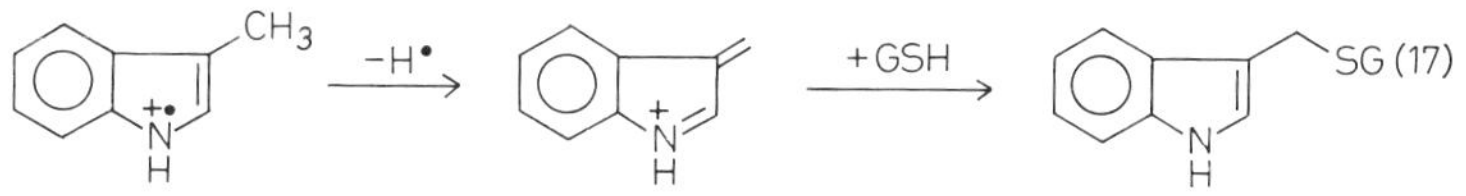

57 58 59

Little or no intermolecular isotope effects have been observed in N-dealkylation reactions. These results are compatible with the rate-limiting oxidation of nitrogen, rather than hydrogen abstraction from carbon, during these reactions [1, 39, 101, 108, 120, 132] studies using aromatic [68, 102] amines are in agreement with observations on N-dealkylations of amine cation radicals generated through oxidation by chlorine dioxide [77], permanganate [115], electromechanically [93] and photochemically [90]. Thus, the larger intramolecular compared to the intermolecular isotope effect is compatible with isotope discrimination in a product determining step from an oxidized nitrogen intermediate such as 52.

VII. Conclusion

Hepatic N-oxidation/reduction reactions are catalyzed primarily by the flavine amine oxidase enzyme (chapter IV) and the P450 system. The flavin amine oxidase enzyme has been purified and many isozymes of P450 have also been purified. Moreover, the isolation of other P450 isozymes, including those from human liver sources as well as the constitutive forms found in various animal species, occupies the efforts of a number of laboratories. Consequently, studies directed at unequivocally demonstrating the ability of these enzymes to catalyze various reactions are becoming feasible.

A dilemma arises, however, as purer forms of these enzymes are used. The dilemma is that, although the pure enzymes can be used to unequivocally demonstrate that these enzymes *can* catalyze a particular reaction, the significance of this metabolism, in vivo, cannot be determined. Future research will be required to bridge this gap.

Much of the mutagenicity and carcinogenicity of aromatic amines can be traced to the two-electron oxidation of these compounds. N-Hydroxylation is usually proposed to be the essential biotransformation in this process since this could activate the N-atom towards reaction with critical biological nucleophiles. The loss of hydroxide ion during this displacement reaction is, however, energetically unfavorable. Moreover, since protonation of the amine nitrogen is more favorable than protonation of the oxygen, the loss of water is also unlikely. Consequently, the activation of nitrogen, solely through N-hydroxylation, is questionable. N-Hydroxyacetaminophen, for example, is quite stable to reactions with various nucleophiles [Nelson, personal communication]. A better leaving group is usually required for N-activation, such as the formation of N-sulfate of acetylaminofluorene [99].

Examples are also available, however, that suggest that N-hydroxylated compounds are directly reactive. Chemical precedence exists [124] from the addition of bisulphite to phenyl hydroxylamine to yield substituted anilines. Several examples have been suggested for the reactivity of aryl hydroxylamines with biological nucleophiles. For example, the N-hydroxylated glutamic acid pyrolysate product has been shown to be directly mutagenic [80] and Ortiz de Montellano et al. [110] have suggested that the N-hydroxy metabolite of aminobenztriazole spontaneously decomposes with the loss of two molecules of N_2 and benzyne. In addition microsomal reduction of nitro compounds to hydroxylamines occurs [123]. Recent stoichiometry studies on the reduction of 5-nitroimidazoles by Miwa et al. [100] suggest that an electrophilic hydroxylamine intermediate is formed that can alkylate proteins and may be responsible for the genotoxic activity of these agents [139]. This subject will certainly benefit from future studies.

Finally, there are very few chemical model systems for P450 N-oxidation/reduction reactions. The elucidation of chemical mechanisms for these P450-catalyzed reactions would benefit from a greater effort in developing suitable chemical model systems for these reactions.

Acknowledgments

The authors are indebted to Dr. Jon Fukuto for helpful discussions and information on nitrogen oxidation mechanisms and to Mrs. Terry Rafferty for assistance in the preparation of the manuscript.

References

1 Abdel-Monem, M.M.: Isotope effects in enzymatic N-demethylation of tertiary amines. J. med. Chem. *18:* 427–430 (1975).
2 Augusto, O.; Beilan, H.S.; Ortiz de Montellano, P.R.: The catalytic mechanism of cytochrome P450. Spin-trapping evidence for one-electron substrate oxidation. J. biol. Chem. *257:* 11288–11295 (1982).
3 Beckett, A.H.: Metabolic oxidation of aliphatic basic nitrogen atoms and their alpha-carbon atoms. Xenobiotica *1:* 365–383 (1971).
4 Beckett, A.H.; Belanger, P.M.: Idetification of three metabolic products of phentermine after liver microsomal incubation. Xenobiotica *4:* 509–519 (1974).
5 Beckett, A.H.; Belanger, P.M.: The mechanism of metabolic N-oxidation of phentermine and chlorphentermine to their hydroxylamino- and nitroso-compounds. J. Pharm. Pharmacol. *26:* 558–560 (1974).
6 Beckett, A.H.; Belanger, P.M.: The microsomal N-oxidation of phentermine. J. Pharm. Pharmacol. *28:* 692–699 (1976).
7 Beckett, A.H.; Gorrod J.W.; Jenner, P.: Absorption of (--)-nicotine-4′oxide in man and its reduction in the gastrointestinal tract. J. Pharm. Pharmacol. *22:* 722–723 (1970).
8 Beckett, A.H.; Jones, G.R.: In vitro metabolic studies using the metabonates of amphetamine, 2-nitroso-1-phenylpropane, and benzyl methyl ketoxime. J. Pharm. Pharmacol. *27:* 60P (1975).
9 Beckett A.H.; Jones, G.R.: Metabolic oxidation of aralkyl oximes to nitro compounds by fortified 9000-g liver supernatants from various species. J. Pharm. Pharmacol. *29:* 416–421 (1977).
10 Belvedere, G.; Rovei, V.; Pantarotto, C.; Frigerio, A.: Mass spectrometric identification of an N-oxide formed by incubation of *N,N*-dimethyl-^{5}H-dibenzo(a,d)cycloheptene-delta 5, gamma-propylamine with rat liver microsomes. Biomed. Mass Spec. *1:* 329–331 (1974).
11 Berzini, A.P.; Traylor, T.G.: NMR studies of P-450 model systems: new structural probes for sulfur-containing hemoproteins. Biochem. Biophys. Res. Commun. *87:* 229–235 (1979).

12 Bickel, M.H.: The pharmacology and biochemistry of N-oxides. Pharmacol. Rev. *21:* 325–355 (1969).
13 Bickel, M.H. Liver metabolic reactions: tertiary amine N-dealkylation, tertiary amine N-oxidation, N-oxide reduction, and N-oxide N-dealkylation. I. Tricyclic tertiary amine drugs. Archs. Biochem. Biophys. *148:* 54–62 (1972).
14 Burka, L.T.; Guengerich, F.P.; Willard, R.J.; MacDonald, T.L.: Mechanism of cytochrome P450 catalysis. Mechanism of N-dealkylation and amine oxide deoxygenation. J. Am. chem. Soc. *107:* 2549–2551 (1985).
15 Chang, C.K.; Dolphin, D.: Ferrous porphyrin-mercaptide complexes. Models for reduced cytochrome P-450. J. Am. chem. Soc. *97:* 5948–5950 (1975).
16 Chang, C.K.; Dolphin, D.: Oxygen binding to mercaptide-heme complexes. Models for reduced cytochrome P-450. J. chem. Soc. *98:* 1607–1609 (1976).
17 Chang, C.K.; Kuo, C.K.: Reaction of iron (III) porphyrins and iodosoxylene. The active oxene complex of cytochrome P-450. J. Am. chem. Soc. *101:* 3413–3414 (1979).
18 Chang, C.K.; Traylor, T.G.: Proximal base influence on the binding of oxygen and carbon monoxide to heme. J. Am. chem. Soc. *95:* 8477–8479 [1973].
19 Cho, A.K., Lindeke, B.; Hodshon, B.J.: The N-hydroxylation of phentermine (2-methyl-1-phynylisopropylamine) by rabbit liver microsomes. Res. Commun. chem. Pathol. Pharmacol. *4:* 519–527 (1972).
20 Cho, A.K.; Sum, C.Y.; Jonsson, J.; Lindeke, B.: The role of N-hydroxylation in the metabolism of phentermine and amphetamine by liver preparations; biological oxidations of nitrogen, pp. 15–23 (Elsevier/North-Holland Biomedical Press, Amsterdam 1978).
21 Collman, J.P.: Synthetic models for the oxygen-binding hemoproteins. Accts. Chem. Res. *10:* 265–272 (1977).
22 Collman, J.P.; Hegedus, L.S.: Dioxygen and its relatives – survey of organotransition-metal complex according to ligand; in principles and applications of organo transition metal chemistry, pp. 154–155 (University Science Books, Mill Valley 1980).
23 Collman, J.P.; Sorrell, T.N.: A model for the carbonyl adduct of ferrous cytochrome P450. J. Am. chem. soc. *97:* 4133–4134 (1975).
24 Cooper, D.Y.; Levin, S.; Narasimhulu, S.; Rosenthal, O.; Estabrook, R.W.: Photochemical action spectrum of the terminal oxidase of mixed function oxidase systems. Science *147:* 400–402 (1965).
25 Coutts, R.T.; Beckett, A.H.: Metabolic N-oxidation of primary and secondary aliphatic medicinal amines. Drug Metab. Rev. *6:* 51–104 (1977).
26 Coutts, R.T.; Dawe, R.; Dawson, G.W.; Kovach, S.H.: In vitro metabolism of 1-phenyl-2-propanone oxime in rat liver homogenates. Drug. Metab. Disposit. *4:* 35–39 (1976).
27 Cramer, J.W.; Miller, J.A.; Miller, E.C.: N-Hydroxylation: a new metabolic reaction observed in the rat with the carcinogen 2-acetylaminofluorene. J. biol. Chem. *235:* 885–888 (1960).
28 Cucinell, S.A.; Cesraili, Z.H.; Dayton, P.G.: Microsomal N-oxidation of dapson as a cause of methemoglobin formation in human red cells. Am. J. trop. Med. Hyg. *21:* 322–331 (1972).
29 Dagne, E.; Castagnoli, N.: Cotinine N-oxide, a new metabolite of nicotine. J. med. Chem. *15:* 840–841 (1972).

30 Dajani, R.M.; Gorrod, J.W.; Beckett, A.H.: In vitro hepatic and extra-hepatic reduction of (minus)-nicotine-1′-N-oxide in rats. Biochem. Pharmacol. *14:* 109–117 (1975).

31 Dawson, J.H.; Holm, R.H.; Trudell, J.R.; Barth, G.; Linder, R.E.; Bunnenberg, E.; Djerassi, C.; Tang, S.C.: Oxidized cytochrome P-450. Magnetic circular dichroism evidence for thiolate ligation in the substrate-bound form. Implications for the catalytic mechanism. J. Am. chem. Soc. *98:* 3707–3708 (1976).

32 Dicken, C.M.; Lu, F.L.; Nee, M.W.; Bruice, T.C.: Kinetics and mechanisms of oxygen transfer in the reaction of *p*-cyano-*N,N*-dimethylaniline-N-oxide with metalloporphyrin salts. 2. Amine oxidation and oxygen transfer to hydrocarbon substrates accompanying the reaction of *p*-cyano-*N,N*-dimethylaniline with meso-(tetraphenylporphinato)iron (III) chloride. J. Am. chem. Soc. *107:* 5775–5789 (1985).

33 Dignan, J.D.; Strobel, H.W.: NADPH-Cytochrome P-450 reductase from rat liver: purification by affinity chromatography and characterization. Biochemistry *16:* 1116–1123 (1977).

34 Dolphin, D.; James, B.R.: Metalloporphyrins. Catalysts for dioxygen reduction and P-450-type hydroxylations. ACS Symp. Ser. *211:* 99–115 (1983).

35 Duncan, J.D.; Cho, A.K.: N-Oxidation of phentermine to N-hydroxyphentermine by a reconstituted cytochrome P-450 oxidase system from rabbit liver. Mol. Pharmacol. *22:* 235–238 (1982).

36 Duncan, J.D.; Distefano, E.W.; Miwa, G.T.; Cho, A.K.: Role of superoxide in the N-oxidation of N-(2-methyl-1-phenyl-2-propyl) hydroxylamine by the rat liver cytochrome P-450 system. Biochemistry *24:* 4155–4161 (1985).

37 Dybing, E.; Aune, T.; Nelson, S.D.: Metabolic activation of 2,4-diaminoanisole, a hair-dye component. II. Role of cytochrome P-450 metabolism in irreversible binding in vitro. Biochem. Pharmacol. *28:* 43–50 (1979).

38 Foster, A.B.; Griggs, L.J.; Jarman, M.; Van Maanen, J.M.S.; Schulten, H.R.: Metabolism of Tamoxifen by rat liver microsomes: formation of the N-oxide, a new metabolite. Biochem. Pharmacol. *29:* 1977–1979 (1980).

39 Foster, A.B.; Jarman, M.; Stevens, J.D.; Thomas, P.; Westwood, J.H.: Isotope effects in O- and N-demethylations mediated by rat liver microsomes: an application of direct insertion electron impact mass spectrometry. Chem.-Biol. Interact. *9:* 327–340 (1974).

40 French, J.S.; Coon, M.J.: Properties of NADPH-cytochrome P-450 reductase purified from rabbit liver microsomes. Archs. Biochem. Biophys. *195:* 565–577 (1979).

41 Frigario, A.; Negrini, P.; Cappellini, L.; Rotilio, D.; De Pascale, A.; Rossi, E.: In vitro studies on the metabolism of dibenzo (c,f)-(1,2)diazepine. Eur. J. Drug Metab. Pharmacokinet. *1:* 35–41 (1979).

42 Fujita, S.; Okada, Y.; Peisach, J.: Inhibition of azoreductase activity by antibodies against cytochromes P-450 and P-448. Biochem. Biophys. Res. Commun. *102:* 492–498 (1981).

43 Fujita, S.; Peisach, J.: Liver microsomal cytochromes P-450 and azoreductase activity. J. biol. Chem. *253:* 4512–4513 (1978).

44 Gal J.; Gruenke, L.D.; Catagnoli, N., Jr.: N-Hydroxylation of 1-(2,5-demethoxy-4-methylphenyl)-2-aminopropane by rabbit liver microsomes. J. med. Chem. *18:* 683–688 (1975).

45 Gal, J.; Wright, J.; Cho, A.K.: In vitro metabolism of amphetamine: an apparent enantiomeric interaction. Res. Commun. chem. pathol. Pharmacol. *15:* 525–540 (1976).
46 Gillette, J.R.; Kamm, J.J.; Sasame, H.: Mechanism of *p*-nitrobenzoate reduction in liver: the possible role of cytochrome P-450 in liver microsomes. Mol. Pharmacol. *4:* 541–548 (1968).
47 Gorrod, J.W.: Biological oxidation of aromatic heterocyclic amines. Xenobiotica *1:* 349–359 (1971).
48 Gorrod, J.W.: Differentiation of various types of biological oxidation of nitrogen in organic compounds. Chem.-Biol. Interact. *289:* 289–303 (1973).
49 Gorrod, J.W.; Damani, L.A.: Effect of various potential inhibitors, activators and inducers on the N-oxidation of 3-substituted pyridines in vitro. Xenobiotica *9:* 219–226 (1979).
50 Gorrod, J.W.; Temple, D.J.; Beckett, A.H.: The differentiation of N-oxidation and N-dealkylation of N-ethyl-N-methylaniline by rabbit liver microsomes as distinct metabolic routes. Xenobiotica *5:* 465–474 (1975).
51 Groves, J.T.: Key elements of chemistry of cytochrome-P-450. The oxygen rebound mechanism. J. chem. Ed. *62:* 928–931 (1985).
52 Groves, J.T.; Kruper, W.J.: Preparation and characterization of an oxoporphinatochromium (V) complex. J. Am. chem. Soc. *101:* 7613–7615 (1979).
53 Groves, J.T.: Kruper, W.J: Metalloporphyrins in oxidative catalysis – oxygen-transfer reactions of oxochromium porphyrins. Israel J. Chem. *25:* 148–154 (1985).
54 Groves, J.T.; McClusky, G.A.: Aliphatic hydroxylation via oxygen rebound. Oxygen transfer catalyzed by iron. J. Am. chem. Soc. *98:* 859–861 (1976).
55 Groves, J.T.; McClusky, G.A.; White, R.E.; Coon, M.J.: Aliphatic hydroxylation by highly purified liver microsomal cytochrome P-450. Evidence for a carbon radical intermediate. Biochem. biophys. Res. Commun. *81:* 154–160 (1978).
56 Groves, J.T.; Nemo, T.E.: Aliphatic hydroxylation catalyzed by iron porphyrin complexes. J. Am. chem. Soc. *105:* 6243–6248 (1983).
57 Groves, J.T.; Nemo, T.E.: Epoxidation reactions catalyzed by iron porphyrin oxygen transfer from iodosyl benzene. J. Am. chem. Soc. *105:* 5786–5791 (1983).
58 Groves, J.T.; Nemo, T.E.; Myers, R.S.: Hydroxylation and epoxidation catalyzed by iron-porphine complexes. Oxygen transfer from iodosylbenzene. J. Am. chem. Soc. *101:* 1032–1033 (1979).
59 Haber, F.; Weiss, J.: The catalytic decomposition of hydrogen peroxide by iron salts. Proc. R. Soc. ser. A *147:* 332–333 (1934).
60 Hamilton, G.A.: Chemical models and mechanisms for oxygenases; in Molecular mechanisms of oxygen activation, pp. 405–451 (Academic Press, New York 1974).
61 Hanzlik, R.P.; Tullman, R.H.: Suicidal inactivation of cytochrome P-450 by cyclopropylamines. Evidence for cation-radical intermediates. J. Am. chem. Soc. *104:* 2048–2050 (1982).
62 Harada N.; Omura, T.: Participation of cytochrome P-450 in the reduction of nitro compounds by rat liver microsomes. J. Biochem. *87:* 1539–1554 (1980).
63 Hashimoto, Y.; Shudo, K.; Okamoto, T.: Activation of a mutagen, 3-amino-1-methyl-5H-pyrido[4,3-b]indole. Identification of 3-hydroxyamino-1-methyl-5H-pyrido[4,3-b]indole and its reaction with DNA. Biochem. Biophys. Res. Commun. *96:* 355–362 (1980).

64 Hashimoto, Y.; Takeda, K.; Shudo, K.; Okamoto, T.; Sugimura, T.; Kosuge, T.: 3-Amino-2-methyl-5h-pyrido[4,3-b]indole, a potent mutagen isolated from tryptophan pyrolysate. Chem.-Biol. Interact. *23:* 137–140 (1978).

65 Hata, Y.; Watanabe, M.; Matsubara, T.; Touchi, A.: Fragmentation reaction of ylide. 5. A new metabolic reaction of aziridine derivatives. J. Am. chem. So. *98:* 6033–6036 (1976).

66 Hata, Y.; Watanabe, M.; Matsubara, T.; Touchi, A.: New metabolic reaction of nitrogen mustards. Chem. Lett. *10*: 1069–1072 (1978).

67 Hecht, S.S.; Chen C.H.; Hoffmann, D.: Metabolic beta-hydroxylation and N-oxidation of N′-nitrosonornicotine. J. med. Chem. *23:* 1175–1178 (1980).

68 Heimbrook, D.C.; Murray, R.I.; Egeberg, K.D.; Sligar, S.G.: Demethylation of N,N-dimethylaniline and *p*-cyano-N,N-dimethylaniline and their N-oxides by cytochromes $P450_{1m2}$ and $P450_{cam}$. J. Am. chem Soc. *106:* 1514–1515 (1984).

69 Henderson, P.T.; Vree, T.B.; Van Ginneken, C.A.M.; Van Rossum, J.M.: Activation energies of alpha-C-oxidation and N-oxidation of N-alkyl-substituted amphetamines by rat liver microsomes. Xenobiotica *4:* 121–130 (1974).

70 Hernandez, P.M.; Mezel, P.; Gillette, J.R.: Mechanism of action of mammalian hepatic azoreductase. II. Effects of phenobarbital and 3-methylcholanthrene on carbon monoxide sensitive and insensitive azoreductase activities. Biochem. Pharmacol. *16:* 1877–1888 (1967).

71 Hlavica, P.; Hulsmann, S.: Studies on the mechanism of hepatic microsomal N-oxide formation. N-Oxidation of N,N-dimethylaniline by a reconstituted rabbit liver microsomal cytochrome P-448 enzyme system. Biochem. J. *182:* 109–116 (1979).

72 Hlavica, P.; Kehl, M.: Studies of the mechanism of hepatic microsomal N-oxide formation. The role of cytochrome P-450 and mixed-function amine oxidase in the N-oxidation of N,N-dimethylaniline. Biochem. J. *164:* 487–496 (1977).

73 Hlavica P.; Kiese, M.: N-Oxygenation of N-alkyl- and N,N-dialkylanilines by rabbit liver microsomes. Biochem. Pharmacol. *18:* 1501–1509 (1969).

74 Horecker, B.L.: Triphosphopyridine nucleotide-cytochrome reductase in liver. J. biol. Chem. *183:* 593–605 (1950).

75 Hucker, H.B.: Phenylacetone oxime, an intermediate in amphetamine deamination. Drug. Metab. Disposit. *1:* 332–336 (1973).

76 Hucker, H.B.; Michniewicz, B.M.; Rhodes, R.E.: Intermediate in the oxidative deamination of amphetamine. Biochem. Pharmacol. *20:* 2123–2128 (1971).

77 Hull, L.A.; Davis, G.T.; Rosenblatt, P.H.; Williams, H.K.R.; Weglein, R.C.: Oxidations of amines. III. Duality of mechanism in the reaction of amines with chlorine dioxide. J. Am. chem. Soc. *89:* 1163–1170 (1967).

78 Hustedt, G.; Kiese, M.: Umsetzungen von Acetanilid und Acetylphenylhydroxylamin in Organismus. Naunyn-Schmiedebergs Arch. Exp. Path. Pharmak. *236:* 435–448 (1959).

79 Ishii, K.; Yamazoe, Y.; Kamataki, T.; Kato, R.: Metabolic activation of mutagenic tryptophan pyrolysis products by rat liver microsomes. Cancer Res. *40:* 2596–2600 (1980).

80 Ishii, K.; Yamazoe, Y.; Kamataki, T.; Kato, R.: Metabolic activation of glutamic acid pyrolysis products, 2-amino-6-methyldipyrido[1,2-a:3′,2′-d]imidazole and 2-aminodipyrido[1,2--a:3′,2′-d]imidazole, by purified cytochrome P-450. Chem.-Biol. Interact. *38:* 1–13 (1981).

81 Iwasaki, K.; Noguchi, H.; Kato, R.; Imai, Y.; Sato, R.: Reduction of tertiary amine N-oxide by purified cytochrome P-450. Biochem. biophys. Res. Commun. *77:* 1143–1149 (1977).
82 Jenner, P.: The role of nitrogen oxidation in the excretion of drugs and foreign compounds. Xenobiotica *1:* 399–418 (1971).
83 Jenner, P.; Gorrod, J.W.; Beckett, A.H.: Species variation in the metabolism of r−(+)- and S−(−)-nicotine by alpha-C-and N-oxidation in vitro. Xenobiotica *3:* 573–580 (1973).
84 Johnson, E.F.; Levitt, D.S.; Müller-Eberhard, U.; Thorgeirsson, S.S: Catalysis of divergent pathways of 2-acetylaminofluorene metabolism by multiple forms of cytochrome P-450. Cancer Res. *40:* 4456–4459 (1980).
85 Kadlubar, F.F.; Miller, J.A.; Miller, E.C.: Microsomal N-oxidation of the hepatocarcinogen N-methyl-4-aminoazobenzene and the reactivity of N-hydroxy-N-methyl-4-aminoazobenzene. Cancer. Res. *36:* 1196–1206 (1976).
86 Kato, R.; Iwasaki, K.; Noguchi, H.: Stimulatory effect of FMN and methyl viologen on cytochrome P-450 dependent reduction of tertiary amine N-Oxide. Biochem. Biophys. Res. Commun. *72:* 267–273 (1976).
87 Kato, R.; Sugiura, M.: The reduction of tertiary amine N-oxides by hepatic microsomes and mitochondira; in Gorrod Biological oxidations of nitrogen, pp. 461–470 (Elsevier/North Holland, Biomedical Press, Amsterdam 1978).
88 Klingenberg, M.: Pigments of rat liver microsomes. Archs. Biochem. Biophys. *75:* 376–386 (1958).
89 Lee, W.A.; Bruice, T.C.: Oxygen-transfer to iron (III) porphyrins by percarboxylic acids and hydroperoxides. A mechanistic criterion for peroxidase and cytochrome P-450. J. Am. chem. Soc. *107:* 513–514 (1985).
90 Lewis, F.D.; Ho, T.I.: On the selectivity of tertiary amine oxidations. J. Am. chem. Soc. *102:* 1751–1752 (1980).
91 Lindeke, B.: The non- and postenzymatic chemistry of N-oxygenated molecules. Drug Metab. Rev. *13:* 71–121 (1982).
92 Lindeke, B.; Anderson, E.; Lundkvist, G.; Jonsson, U.; Eriksson, S.O.: Autoxidation of N-hydroxyamphetamine and N-hydroxyphentermine. Formation of 2-nitroso-1-phenylpropanes and 1-phenyl-2-propane oxime. Acta pharm. Suecica *12:* 183–198 (1975).
93 Lindsay Smith, J.R.; Masheder, D.: Amine oxidation. IX. Electrochemical oxidation of some tertiary amines. Effect of structure on reactivity. J. chem. Soc. Perkin Trans. *II:* 47–51 (1976).
94 Logan, C.J.; Cotter, F.H.; Page, J.A.: The formation of a wood mercapturic acid during the metabolism of an N-methyl aromatic amine, 4-cyano-N,N-dimethylaniline. Biochem. Pharmacol. *33:* 2345–2346 (1984).
95 MacDonald, T.L.; Zirvi, K.; Burka, L.T.; Peyman, P.; Guengerich, F.P.: Mechanism of cytochrome P-450 inhibition by cyclopropylamines. J. Am. chem. Soc. *204:* 2050–2052 (1982).
96 Mansuy, D.; Fontecave, M.: A heme model system for the reduction of substrates by microsomal cytochrome P-450. Biochem. Biophys. Res. Commun. *104:* 1651–1657 (1982).
97 Matsumoto, R.M.; Cho, A.K.; Conversion of N-hydroxyamphetamine to phenylacetone oxime by rat liver microsomes. Biochem. Pharmacol. *31:* 105–108 (1982).

98 Maynard, M.S.; Cho, A.K.: Oxidation of N-hydroxyphentermine to 2-methyl-2-nitro-1-phenylpropane by liver microsomes. Biochem. Pharmacol. *30:* 1115–1119 (1981).

99 Miller, E.C.; Miller, J.A.: Mechanisms of chemical carcinogenesis. Cancer. *47:* 1055–1064 (1981).

100 Miwa, G.T.; Argenbright, L.S.; Lu, A.Y.H.; Kedderis, G.L.: Studies on the role of nitro group reduction in the chemical activation of ronidazole, a 5-nitroimidazole. Fed. Proc. *45:* 577 (1986).

101 Miwa, G.T.; Garland, W.A.; Hodshon, B.J.; Lu; A.Y.H.; Northrop, D.B.: Kinetic isotope effects in cytochrome P-450-catalyzed oxidation reactions. Intermolecular and intramolecular deuterium isotope effects during the N-demethylation of N,N-dimethylphentermine. J. biol. Chem. *255:* 6049–6054 (1980).

102 Miwa, G.T.; Walsh, J.S.; Kedderis, G.L.; Hollenberg, P.F.: The use of intramolecular isotope effects to distinguish between deprotonation and hydrogen atom abstraction mechanisms in cytochrome P450- and peroxidase-catalyzed N-demethylation reactions. J. biol. Chem. *258:* 14445–14449 (1983).

103 Miwa, G.T.; West, S.B.; Huang, M.T.; Lu, A.Y.H.: Studies on the association of cytochrome P-450 and NADPH-cytochrome *c* reductase during catalysis in a reconstituted hydroxylating system. J. biol. Chem. *254:* 5695–5700 (1979).

104 Miyata, N.; Santa, T.; Hiroke, M.: Chemical studies on drug metabolism 6. Deoxygenation of tertiary amine N-oxides and arene oxides by iron (II) porphyrin as a model of cytochrome P450-dependent reduction. Chem. Pharm. Bull. *32:* 377–380 (1984).

105 Moshima, I.; Ogawa, S.: Nuclear magnetic resonance characterization of compounds I and II of horseradish peroxidase. J. Am. chem. Soc. *100:* 7125–7127 (1978).

106 Murray, K.N.; Watson, G.J.; Chaykin, S.: Catalysis of the direct transfer of oxygen from nicotinamide N-oxide to xanthine by xanthine oxidase. J. biol. Chem. *241:* 4798–4801 (1966).

107 Nanni, E.J., Jr.; Sawyer, D.T.: Superoxide-ion oxidation of hydrophenazines, reduced flavins, hydroxylamine, and related substrates via hydrogenation transfer. J. Am. chem. Soc. *102:* 7591–7593 (1980).

108 Nelson, S.D.; Pohl, L.R.; Trager, W.F.: Primary and secondary deuterium isotope effects in N-deethylation reactions. J. Med. Chem. *18:* 1062–1065 (1975).

109 Nocerini, M.R.; Yost, G.S.; Carlson, J.R.; Liberato, D.J.; Breeze, R.G.: Structures of the glutathione adduct of activated 3-methylindole indicates than an imine methide is the electrophilic intermediate. Drug Metab. Disposit. *13:* 690–694 (1985).

110 Ortiz de Montellano, P.R.; Mathews, J.M.; Langry, K.D.: Autocatalytic inactivation of cytochrome P450 and chloroperoxidase by 1-aminobenzotriazole and other aryne precursors. Tetrahedron *40:* 511–519 (1984).

111 Parli, C.J.; Wang, N.; McMahon, R.E.: Enzymic N-hydroxylation of an imine. New cytochrome P-450 dependent reaction catalyzed by hepatic microsomal monooxygenases. J. biol. Chem. *246:* 6952–6955 (1971).

112 Poulos, T.L.; Finzel, B.C.; Gunsalus, I.C.; Wagner, G.C.; Kraut, J.: The 2.6Å crystal structure of *Pseudomonas putida* cytochrome P-450. J. biol. Chem. *260:* 6122–6130 (1985).

113 Powell, M.F.; Pai, E.F.; Bruice, T.C.: Study of tetraphenylporphinato manganese III. Catalyzed epoxidation and demethylation using *p*-cyano-N,N-dimethylaniline-

N-oxide as oxygen donor in a homogeneous system. Kinetics, radiochemical ligation studies and reaction mechanism for a model of cytochrome P450. J. Am. chem. Soc. *106:* 3277–3285 (1984).
114 Powis, G.; Wincentsen, L.: Pyridine nucleotide cofactor requirements of indicine N-oxide reduction by hepatic microsomal cytochrome P-450. Biochem. Pharmacol. *29:* 347–351 (1980).
115 Rosenblatt, P.H.; Davis, G.T.; Hull, L.A.; Forberg, G.D.: Oxidations of amines. IV. Duality of mechanism in the reaction of aliphatic amines with permanganate. J. org. Chem. *33:* 1649–1650 (1968).
116 Sakurai, H.; Yoshimura, T.: Models for coordination site of cytochrome P450 characterization of hemin-thiolato complexes with S, O and N donor ligands by electronic absorption and ESR spectra. J. inorg. Biochem. *24:* 75–96 (1985).
117 Santa, T.; Miyata, N.; Hirobe, M.: Chemical studies on drug metabolism 7. Oxidation of tertiary amines and sulfides by an iron (III) porphyrin-oxygen-sodium dithionite system as a model of cytochrome P450. Chem. Pharm. Bull. *32:* 1252–1255 (1984).
118 Shannon, P.; Bruice, T.C.: A novel P-450 model system for the N-dealkylation reaction. J. Am. chem. Soc. *103:* 4580–4582 (1981).
119 Sharrock, M.; Munek, E.; Debrunner, P.G.; Marshall, V.; Lipscomb, J.O.; Gunsalus, I.G.: Mössbauer studies of cytochrome-$P450_{cam}$. Biochemistry *12:* 258–265 (1973).
120 Shono, T.; Toda, T.; Oshino, N.: Electron transfer from nitrogen in microsomal oxidation of amine and amide. Simulation of microsomal oxidation by anodic oxidation. J. Am. chem. Soc. *104:* 2639–2641 (1982).
121 Sligar, S.G., Kennedy, K.A.; Person, D.C.: Chemical mechanisms for cytochrome P-450 hydroxylation: evidence for acylation of heme-bound dioxygen. Proc. Natn. Acad. Sci. USA *77:* 1240–1244 (1980).
122 Smegal, J.A.; Hill, C.L.: Hydrocarbon functionalization by the iodosylbenzene manganese IV porphyrin complexes from the tetraphenylporphinato manganese III iodosylbenzene catalytic hydrocarbon oxidation system. Mechanism and reaction chemistry. J. Am. chem. Soc. *105:* 3515–3521 (1983).
123 Sternson, L.A.: Detection of arylhydroxylamines as intermediates in the metabolic reduction of nitro compounds. Experientia *31:* 268–270 (1975).
124 Sternson, L.A.; Dixit, A.S.; Becker, A.R.: Reaction of phenylhydroxylamine with bisulfite. A possible model for amine-mediated carcinogenesis. J. org. Chem. *48:* 57–60 (1983).
125 Sugiura, M.; Iwasaki, K.; Kato, R.: Reduction of tertiary amine N-oxides by liver microsomal cytochrome P-450. Mol. Pharmacol. *12:* 322–334 (1975).
126 Sugiura, M.; Iwasaki, K.; Kato, R.: reduced nicotinamide adenine dinucleotide-dependent reduction of tertiary amine N-oxide by liver microsomal cytochrome P-450. Biochem. Pharmacol. *26:* 489–495 (1977).
127 Sum, C.Y.; Cho, A.K.: The N-hydroxylation of phentermine by rat liver microsomes. Drug. metab. Disposit. *5:* 464–468 (1977).
128 Sum, C.Y.; Cho, A.K.: The metabolism of N-hydroxyphentermine by rat liver microsomes. Drug. Metab. Disposit. *7:* 65–69 (1979).
129 Tabushi, I.; Kodera, M.; Yokoyama, M.: Kinetics and mechanism of reductive dioxygen activation catalyzed by the P450 model system iron picket fence as a catalytic centre. J. Am. chem. Soc. *107:* 4466–4473 (1985).

130 Takeuchi, K.J.; Busch, D.H.; Alcock, N.: Guest-host association by transition – metal complexes containing permanent voids – progress toward models for the ternary complex of cytochrome P-450. J. Am. chem. Soc. *103:* 2421–2422 (1981).

131 Tatsumi, K.; Kitamura, S.; Yoshimura, H.: Reduction of nitrofuran derivatives by xanthine oxidase and microsomes. Isolation and identification of reduction products. Archs. Biochem. Biophys. *175:* 131–137 (1976).

132 Thompson, J.A.; Holtzman, J.L.: Deuterium and tritium isotope effects on the microsomal N-demethylation of ethylmorphine. Drug. metab. Disposit. *2:* 577–582 (1974).

133 Torrielli, M.V.; Slater, T.F.: Inhibition of NADPH-cytochrome *c* reductase by propyl gallate. Biochem. Pharmacol. *20:* 2027–2032 (1971).

134 Uehleke, H.: The role of cytochrome P-450 in the N-oxidation of individual amines. Drug. metab. Disposit. *1:* 299–313 (1973).

135 Uehleke, H.; Breyer, U.; Budczies, B.; Tabarelli, S.; Hellmer, K.H.: The influence of metyrapone on individual types of microsomal N-oxidations. Z. physiol. chem. *352:* 403–411 (1971).

136 Ullrich, V.; Kuthan, H.: Autooxidation and uncoupling in microsomal monooxygenations; in Gustafsson, Carlstedt-Duke, Mode, Rafter, Biochemistry, biophysics and regulation of cytochrome P450, pp. 267–272 (Elsevier/North-Holland Biomedical Press, Amsterdam 1980).

137 Valentine, J.S.: Dioxygen ligand in mononuclear group III transition metal complexes. Chem. Rev. *73:* 235–245 (1973).

138 Vermilion, J.L.; Ballou, D.; Massey, V.; Coon, M.J.: Separate roles for FMN and FAD in catalysis by liver microsomal NADPH-cytochrome P-450 reductase. J. biol. Chem. *256:* 266–277 (1981).

139 Walsh, J.S.; Wang, J.R.; Bagan, E.; Wang, C.C.; Wislocki, P.; Miwa, G.T.: Structural alterations that differentially affect the mutagenic and antitrichomonal activities of 5-nitroimidazoles. J. med. Chem. *30:* 150–156 (1987).

140 Wang, C.Y.; Behrens, B.C.; Ichikawa, M.; Bryan, G.T.: Nitroreduction of 5-nitrofuran derivatives by rat liver xanthine oxidase and reduced nicotinamide adenine dinucleotide phosphate-cytochrome *c* reductase. Biochem. Pharmacol. *23:* 3395–3404 (1974).

141 Weinkam, R.J.; Shiba, D.A.: Metabolic activation of procarbazine. Life Sci. *22:* 937–946 (1978).

142 Weisburger, J.H.; Weisburger, E.K.: Biochemical formation and pharmacological, toxicological, and pathological properties of hydroxylamines and hydroxamic acids. Pharmacol. Rev. *25:* 1–66 (1973).

143 Werringloer, J.; Kawano, S.: The control of the cyclic function of liver microsomal cytochrome P-450: counterpoise – regulation of the electron transfer reactions required for the activation of molecular oxygen; in Gustafsson, Carlstedt-Duke, Mode, Rafter, Biochemistry, biophysics and regulation of cytochrome P450, (Elsevier/North Holland Biomedical Press, Amsterdam 1980).

144 White, R.; Coon, M.J.: Oxygen activation by cytochrome P-450. A. Rev. Biochem. *49:* 315–356 (1980).

145 White, R.E.; Sligar, S.G.; Coon, M.J.: Evidence for a homolytic mechanism of peroxide oxygen-oxygen bond cleavage during substrate hydroxylation by cytochrome P-450. J. biol. Chem. *255:* 11108–11111 (1980).

146 Willi, P.; Bickel, M.H.: Liver metabolic reactions: tertiary amine N-dealkylation, tertiary amine N-oxidation, N-oxide reduction, and N-oxide N-dealkylation. Archs. Biochem. Biophys. *156:* 772–779 (1973).
147 Wislocki, P.; Miwa, G.; Lu; A.Y.H.: Reactions catalyzed by the cytochrome P-450 system; in Jacoby, Enzymatic basis of detoxification, vol. 1, pp. 135–182 (Academic Press, New York 1980).
148 Yang, C.S.: The organization and interaction of monoxygenase enzymes in the microsomal membrane. Life Sci. *21:* 1047–1058 (1977).
149 Yasukochi, Y.; Masters, B.S.S.: Some properties of a detergent-solubilized NADPH-cytochrome *c* (cytochrome P-450) reductase purified by biospecific affinity chromatography. J. biol. Chem. *251:* 5337–5344 (1976).
150 Ziegler, D.M.: Molecular basis for N-oxygenation of sec- and tert-amines; in Gorrod, Damani, The biological oxidation of nitrogen in organic molecules, pp. 43–52 (Horwood, Chichester 1985).
151 Ziegler, D.M.; Pettit, F.H.: Microsomal Oxidases. I. The isolation and dialkylarylamine oxygenase activity of pork liver microsomes. Biochemistry *5:* 2932–2938 (1966).

G.T. Miwa, PhD, Department of Drug Metabolism, Glaxo Inc.,
Research Triangle Park, NC 27709 (USA)

Prog. bas. clin. Pharmac., vol. 1, pp. 63–102 (Karger, Basel 1988)

III. Nitrogenous Compounds as Ligands to Hemoporphyrins – The Concept of Metabolic-Intermediary Complexes

B. Lindeke[a], *U. Paulsen-Sörman*[b]

[a]Medical Department, ACO Lakemedel AB, Solna; and
[b]Pharmacia AB, The Health Care Group, Uppsala, Sweden

I. Introduction

The details of the biochemistry of cytochrome P450 and its proposed mechanism of action with respect to the metabolism of nitrogenous compounds have been described in chapter II. This chapter focuses on the interaction of certain electron-rich nitrogenous substrates with the cytochrome P450 enzyme system to form spectrally defined complexes. These complexes have different stabilities and some of them can inhibit the oxygenase activity of the enzyme. In this presentation the spectral properties of hemoporphyrins in general and cytochrome P450 in particular are discussed, followed by a discussion of the properties of the heme iron. Then the characteristics of complex formation from different types of nitrogen functionalities are described. Finally, structure activity relationships will be discussed for some nitrogen functionalities and possible mechanisms leading to the formation of metabolic intermediary complexes will be considered.

II. Spectral Perturbations

The inherent chromophoric properties of protoheme (fig. 1a) – which, apart from chlorophyll, is the most common naturally occurring porphyrin derivative – have made spectrophotometric methods [11] a useful tool in analyzing proteins containing this hemoporphyrin. It con-

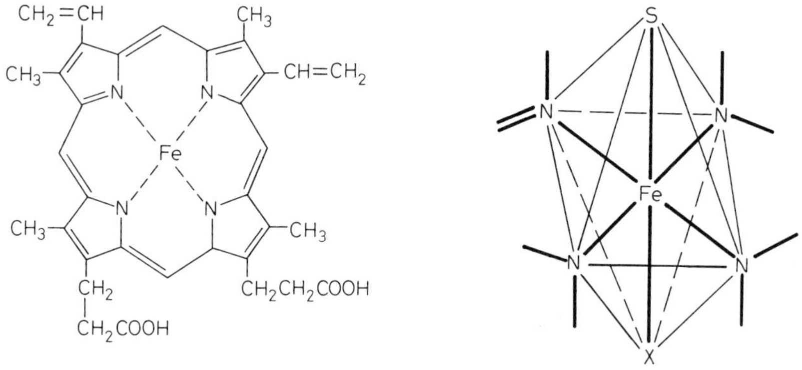

a *b*

Fig. 1. a The chemical structure of protoheme. *b* The six octahedral bonds of iron in the protoheme.

stitutes the prosthetic group of hemoglobin, myoglobin, catalase, cytochromes and many other proteins. A pronounced spectral characteristic of all porphyrin derivatives is that upon excitation they give rise to an intense $\pi \rightarrow \pi^*$ transition at about 400 nm (the Soret band), which is due to delocalization extending throughout the macrocycle, and several other characteristic longer wavelength bands. The spectral characteristics are sensitive to changes in the hemoporphyrin but also in its environment. Thus, changes in the oxidation (Fe^{3+} or Fe^{2+}) or spin (ligation) states of the iron atom can be distinguished as can conformational changes in the protein surrounding the heme group.

One way in which changes in the environment of the chromophore can be demonstrated is by difference spectroscopy [11, 19] whereby the difference between the chromophore in two environments is directly measured. In studies of cytochrome-P450-mediated reactions mainly two types of spectral perturbations in the Soret region, classified as type I and type II spectral changes [65, 77, 95], have received special interest. Spectrally, the two types of perturbations are essentially mirror images of each other and result from the interaction of the enzyme in its oxidized (Fe^{3+}) form with various substrates (fig. 2a). Substrates (exogenous as well as endogenous) have been classified as type I or type II depending on their ability to cause either type I or type II spectral changes upon addition to the enzyme. Type I spectral changes have been considered to be associated with metabolism, i.e. to be the optical manifestation of the binding of substrate to the protein moiety of the enzyme [95] – while type

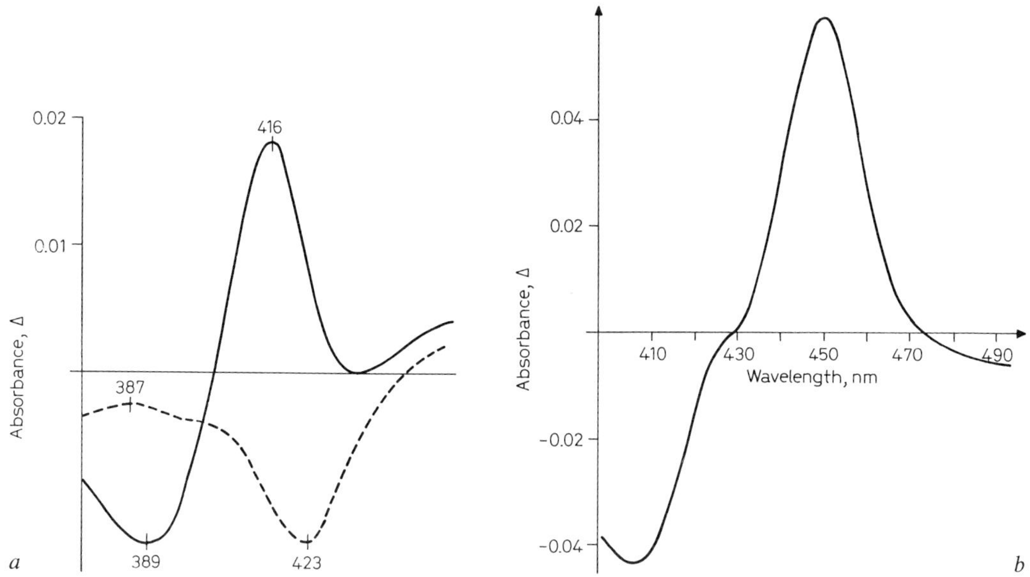

Fig. 2. a Aerobic difference spectra of liver microsomes with the iron in its oxidized (Fe^{3+}) state, in the presence of N-hydroxyamphetamine (— = type II) and N-hydroxyphentermine (--- = type I). *b* Difference spectra of liver microsomes with the iron in its reduced (Fe^{2+}) state, in the presence of carbon monoxide.

II spectral perturbations are believed to reflect a direct interaction between the substrate and the heme iron (ligand binding) [95]. However, there are also examples of substrates which can interact to produce both types of spectral changes, and much remains to be understood about the mechanisms of these changes.

The most significant property of the hemoporphyrins is their capacity to bind molecular oxygen which is the basis of oxygen transportation by hemoglobin as well as the activation of molecular oxygen by various cytochromes. However, oxygen binding can occur only when the iron of the heme is in its reduced (Fe^{2+}) state, and in cytochrome P450-mediated reactions, reduction of the hemin iron of the enzyme-substrate complex generates a high-spin (pentacoordinated) ferrous form which is capable of binding oxygen. Upon binding of oxygen, cytochrome P450 forms a low-spin (hexacoordinated) ferrous enzyme-substrate oxygen complex [89] (see section I.B.).

Carbon monoxide (CO) is notable among compounds besides oxygen that can act as ligands to the reduced iron of protoheme. The Fe^{2+}

carbonyl complexes of the cytochrome P450 enzymes exhibit difference spectra with a maximum around 450 nm, and this band is the basis for their classification as cytochrome P450 pigments (fig. 2b) [76]. Similar absorbance difference spectra of the CO-complexes of hemoglobin and myoglobin have maxima at around 420 nm.

III. Some General Properties of Hemin Iron

Iron can form six octahedral bonds by hybridization of its 4 s, 4 p and 3 d orbitals [79]. Thus, in addition to the four pyrrole nitrogens that fix iron in the porphyrin (fig. 1a), two additional ligands can be accommodated on the iron (fig. 1b). In the cytochrome P450 enzymes, the fifth ligand is believed to be a mercaptide anion [52] or an imidazole nitrogen [89]. The sixth ligand is exchangeable and is proposed to be a hydroxyl group [57], or water [89] when the enzymes appear in their low-spin Fe^{3+} substrate-free form [89, 97]. Following substrate binding and reduction of the iron, O_2 is substituted for the sixth ligand. It is evident (fig. 3) that any entity other than O_2 which can bind to the heme iron will obstruct O_2 binding and in the case of the cytochrome P450 enzymes will inhibit the oxygenase capacity of the enzymes. Thus, ligands interfering with the high-spin (pentacoordinated) Fe^{3+} state will reverse the transformation of the heme into its low-spin (hexacoordinated) Fe^{3+} state and prevent the reduction of the enzyme-substrate complex. Accommodation of ligands other than O_2 at the high-spin (pentacoordinated) Fe^{2+} state will generate ternary low-spin (hexacoordinated) Fe^{2+} complex devoid of O_2 and the monooxygenase activity.

There are a vast number of entities apart from oxygen that act as ligands to reduced and/or oxidized iron of the protoheme. They range from simple diatomic molecules, such as CO, nitric oxide (NO) and cyanide anion (CN^-), to complex endogenous and xenobiotic molecules. One feature they have in common, though, is the availability of at least one lone-pair of electrons (Lewis bases) or a negative charge. Compounds containing oxygen, sulfur or especially nitrogen atoms act as ligands to oxidized protoheme (Fe^{3+}) which, in cytochrome-P450-mediated reactions, can be manifested by the appearance of the type II spectral changes as previously mentioned (section II). CN^- has a high affinity for oxidized (Fe^{3+}) iron and reacts readily with cytochromes in their ferric state but not with hemoglobin (Fe^{2+}). The toxic effect of

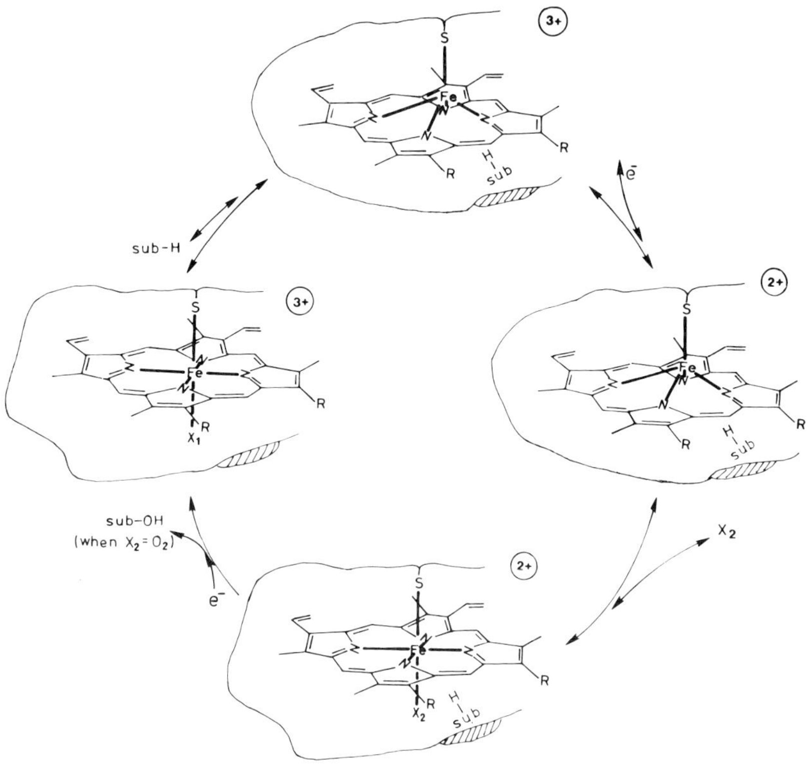

Fig. 3. Schematic reaction cycle of cytochrome P450, showing proposed ligand and spin state changes.

cyanide poisoning is a result of a cytotoxic hypoxia caused by the formation of the cytochrome-oxidase-CN complex [101]. Isocyanides ($R-N\equiv C$) also bind to Fe^{3+} but apparently only in the cytochrome P450 enzymes [107]. Otherwise, this ligand has a pronounced affinity for hemoproteins containing reduced iron (Fe^{2+}). The Fe^{2+}-isocyanide complexes of the cytochrome P450 enzymes exhibit difference spectra very similar to those of the corresponding Fe^{2+}-carbonyl complexes (fig. 2b) but with the maximum appearing at about 455 nm [103].

NO has a high affinity for the reduced iron (Fe^{2+}) of many mammalian hemoporphyrins, such as hemoglobin, myoglobin and various cytochromes [20, 21, 34, 48, 78]. Plant hemes also form these complexes during the six-electron reduction of nitrite to ammonia, a reaction catalyzed by many plant nitrite oxidoreductases that contain sirohemes

[9, 32]. Thus, ligand binding of NO to the iron of hemoporphyrins is a reaction not alien to the biochemistry of the living cell.

NO, an odd electron molecule like oxygen, the nitroxide radical, and organic C-nitroso compounds are isoelectronic with O_2 [79]. Other Lewis base type ligands with affinity for reduced hemoporphyrins include certain trivalent phosphorous compounds (e.g. $[RO]_3P$) [14] and stannous halides (e.g. SnF_2) [13].

Apart from the directly acting ligands, there are numerous examples of cytochrome P450 ligand complexes produced during metabolic transformations of certain classes of xenobiotics.. Franklin [30] has designated the term 'metabolic intermediary complexes' (MI complexes) for this phenomenon. The spectral perturbations associated with cytochrome P450 MI complex-forming reactions are somewhat different from those exerted by the directly acting ligands. Addition of a ligand precursor to the enzyme results in an absorbance change which increases with time (fig. 4) while complex formation with the directly acting ligands is characterized by a fixed absorbance difference (fig. 2b). The presence of the kinetic factor is compatible with the continuous production of a ligand during the metabolism of the substrate in question [30].

Nitrogenous compounds constitute an outstanding group within the class of 'indirectly acting' ligands. This review will focus on ligand formation and the genesis of metabolic intermediary complexes in the metabolism of nitrogenous compounds. The subject has also been discussed by Testa and Jenner [102] in their comprehensive review on cytochrome P450 inhibitors and their mechanism of action.

IV. Nitrogenous Compounds Generating Cytochrome P450 MI Complexes

Benzphetamine was one of the first nitrogenous compounds recognized to generate cytochrome P450 MI complexes. The complexes are identifiable spectrophotometrically, the most striking characteristic in the difference spectrum being a maximum around 456 nm (cf. fig. 4). From the results of their studies, Werringloer and Estabrook [106, 108] concluded that further metabolic conversion of a metabolite of benzphetamine was most likely the cause for the generation of the adduct. Subsequent studies by Franklin [25, 26] revealed that amphetamine and

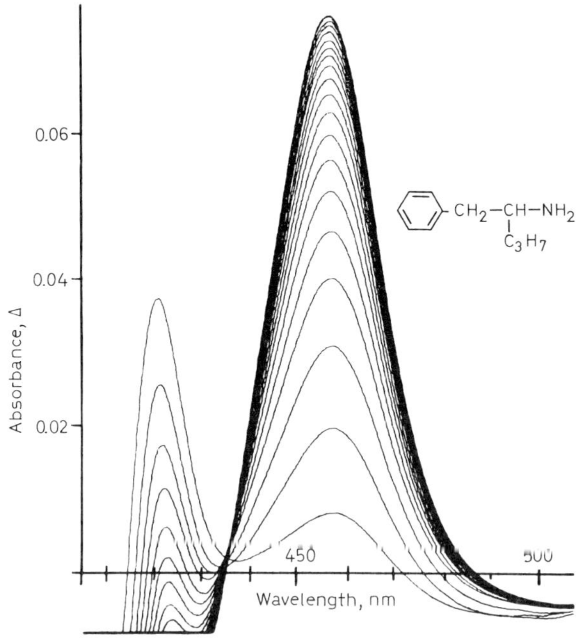

Fig. 4. Difference spectra produced during NADPH-dependent metabolism of α-propyl-substituted 2-phenylethylamine (500 μ*M*). Time difference between each scan was 41 s.

some of its primary, secondary and tertiary amine congeners were also able to generate complexes similar to those obtained from benzphetamine.

An additional finding by Franklin [27] and James and Franklin [44] was that N-hydroxyamphetamine is unique with regard to its potency as a substrate for the complex formation, and other investigators found that complexes can be generated from C-nitroso [51, 66, 68–70] and nitro [66–70, 104] compounds as well. Also apart from the 1-benzylalkylamine-type substrates which are the most thoroughly investigated, complex formation occurs in the metabolism of a number of other amines and nitrogenous compounds. In his review covering work to 1976, Franklin [30] has compiled most compounds then known to generate the cytochrome P450 MI complexes. An updated compilation of nitrogenous cytochrome P450 complex-generating compounds is now presented to include the information gained until the mid-1980s (Appendix, table I).

A. Amines

The ligand for the MI complexes can originate from primary, secondary and tertiary aliphatic as well as aromatic amines. With all types of amines, prerequisites for complex formation are the presence of O_2 and NADPH [30] – i.e. metabolism is necessary and the cytochrome is complexed in the reduced (Fe^{2+}) state (cf. fig. 3). The complex-associated spectral perturbations exhibit maxima in the range 445–457 nm, with lower wavelengths reflecting interactions originating from the aromatic amines [29, 30].

Based on the early findings, Franklin [30] divided the nitrogenous complex-generating substrates into three groups – aliphatic amines related to amphetamine, compounds related to SKF 525-A and aromatic amines. However, while the aromatic amines are distinctly different from the amphetamine group in their interaction with reduced cytochrome P450, newer findings [3, 46] have blurred what was believed previously to be clear-cut differences between the amphetamines and other types of aliphatic amines, such as the SKF 525-A class of compounds. Indeed, despite structural differences, the same mechanisms could apply to both groups (see section VII.B).

A characteristic feature of complex formation from many amines is the prevalence of a lag-phase prior to the commencement of the chromophore generation. Such lag-phases can be quite pronounced, as is the case with amphetamine and benzphetamine [25, 106, 108], but with a series of primary amine analogues of amphetamine they decrease with increasing lipophilicity of the substrate and with increasing substrate concentration [64]. The generation of complexes from many tertiary and secondary amines (e.g. benzphetamine and SKF 525-A) have been shown to exhibit substrate inhibition [3, 12, 28, 30, 35, 46, 64, 106, 108] (see section V.A.).

Finally, it should be realized that there are many amines which have been found inactive in forming MI complexes. Among these, 2-phenylethylamine, phentermine, ephedrine, norephedrine, *p*-hydroxyamphetamine, imipramine and amitriptyline are to be noted.

B. N-Hydroxylamines

All N-hydroxy analogues of MI-complex-generating amines hitherto investigated have been found to generate complexes which are characterized by absorbance maxima appearing at the same wavelengths as those of the corresponding amines (Appendix, table I.C). However,

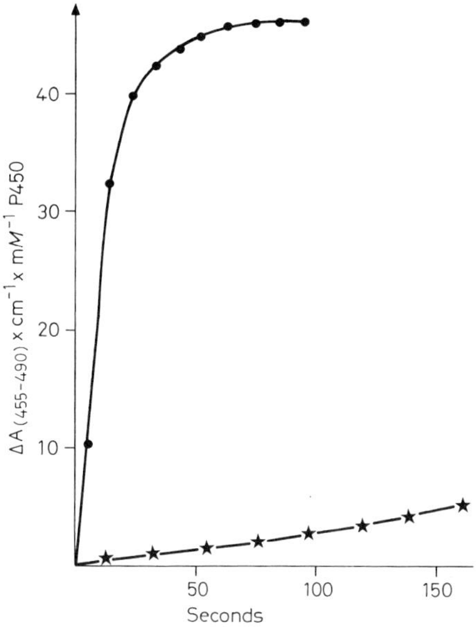

Fig. 5. Cytochrome P450 complex formation as a function of time for α-ethyl-substituted 2-phenylethylamine (★ = lower curve, 50 μ*M*) and the corresponding N-hydroxylamine (● = upper curve, 50 μ*M*).

with the N-hydroxylamines as substrates, the rate of complex formation is generally increased by one to two orders of magnitude, and a considerable increase is seen also in the amount of complex formed (fig. 5) [27, 39, 44, 51, 63, 64, 71]. Moreover, the pronounced lag-phases, noted in studies with certain amines, are never seen with the corresponding N-hydroxylamines.

As with the amines, prerequisites for complex formation with *aliphatic* N-hydroxylamines are the presence of O_2 and NADPH-reduction of the cytochrome. With some easily oxidized *aromatic* N-hydroxylamines, however, the presence of NADPH is not mandatory [70].

The N-hydroxy congeners of 2-phenylethylamine and norephedrine were found to be good substrates for MI complex formation despite the fact that the corresponding amines are inactive [62, 64]. Small and polar hydroxylamines, such as N-methylhydroxylamine, also form complexes [63]. The N-hydroxy analogues of *t*-butyl amines, such as N-hydroxyphentermine, however, remain inactive [25, 51, 71].

C. C-Nitroso Compounds

Complex formation with C-nitroso compounds, the further oxidation products of primary N-hydroxylamines, does occur (Appendix, table I.D) but is obscured by the ease with which these entities undergo chemical conversions [61] (cf. chapter I). A major complication is that synthetic C nitroso compounds are available only as dimers and upon dissolution their dissociation is an equilibrium displaced in favor of the dimer. Furthermore, if α-hydrogens are available oxime formation has to be taken into account. Thus, in the worst case, a solution of C-nitroso compounds could contain three substances (equation 1) – the nitroso dimer (1), the nitroso monomer (2) and the oxime (3) [80].

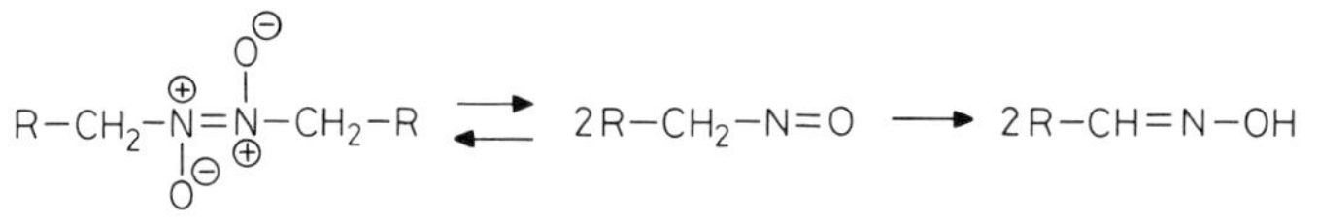

(1) (2) (3)

(1)

As oximes (3) are known not to bind to reduced cytochrome P450 to generate MI complexes [69] (Appendix, table I.D), complex formation with the nitroso species can be assessed although quantitative measurements are difficult.

Complex formation with the C-nitroso compounds differs from that with amines or N-hydroxylamines inasmuch as complexes are generated both by interaction with NADPH-reduced microsomes under aerobic conditions, and by dithionite-reduced microsomes (i.e. anaerobic conditions) [51, 62, 68, 70]. Accordingly, the nitroso compounds are able to form the complexes without the prerequisite of oxidative metabolism. Also, nitrosobenzene has recently [33] been shown to react with cytochrome P450 in the Fe^{3+} form, in the absence of any exogenous reducing agent, to produce the Fe^{2+} complex. This mechanism implicates a spontaneous autoreduction of the iron.

D. Nitrones

Nitrones, the oxidation products of secondary N-hydroxylamines, have been found to form cytochrome P450 complexes (Appendix, table I.E). Presence of O_2 and NADPH-reduction of the cytochrome are absolute requirements.

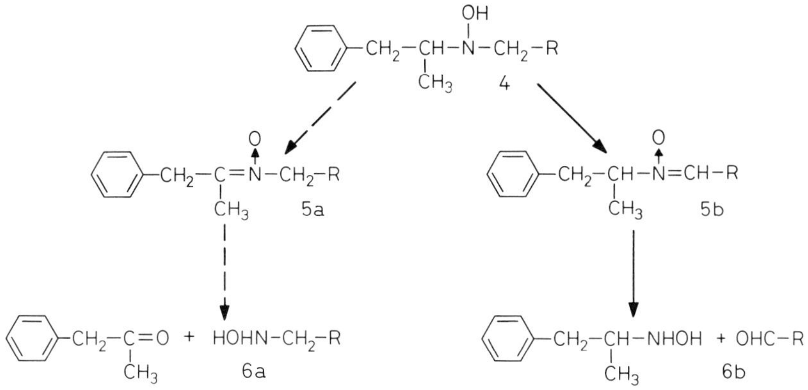

Fig. 6. Possible oxidation routes for N-hydroxymethamphetamine (4, R = H).

It is interesting to note that the two nitrones, 5a and 5b in figure 6, both potential oxidation products of N-hydroxymethamphetamine 4 (R = H), generate complexes with maxima at 453 and 455 nm, respectively. These maxima are identical to complexes obtained from the corresponding primary N-hydroxylamines, 6a and 6b, that constitute the hydrolysis products of 5a and 5b [63].

E. Nitro Compounds

Nitro compounds were examined relatively early in investigations of spectral perturbations associated with substrate interactions with cytochrome P450. Thus, in the early 1970s Ullrich and Schnabel [104] discovered a cytochrome P450 complex with a Soret maximum at 454 nm, upon dithionite reduction of 2-nitropropane in microsomal preparations. Subsequent investigations, especially by Mansuy and coworkers [66–68, 70], rendered generality to the reaction and many nitro compounds have now been found to generate similar complexes upon dithionite reduction (Appendix, table I.E). Although nitro compounds form complexes upon reduction with dithionite, there are no reports whatsoever on complex formation during their microsomal cytochrome P450-mediated metabolism. This would support the notion that reduction of aliphatic nitro compounds is not catalyzed by liver fractions (cf. chapter VI).

Moreover, no complexes are formed during oxidative cytochrome P450-catalyzed denitrification of nitro alkanes [93, 105]. It is thus evident that with nitro compounds reduction of both the substrate and the cytochrome is a prerequisite for complex formation. Dithionite reduces nitro compounds to both the nitroso and the hydroxylamine state (unpublished observations) and simultaneously inflicts anaerobic conditions on the microsomal system.

F. N-Oxides

There are a few reports on MI complex formation attributed to tertiary amine N-oxides (Appendix, table I.E). Thus, a 448-nm absorbance maximum appearing during NADPH-dependent oxidative metabolism of *N,N*-dimethylaniline was considered due to binding of a metabolite formed after N-oxidation of the arylamine [40, 41]. Tiaramide N-oxide, a 1,4-disubstituted piperazine N-oxide, produced a pronounced spectral perturbation with a maximum at 442 nm in rat liver microsomes in the presence of dithionite [99]. Another report [42] describes the formation of similar ferrocytochrome P450 product adducts, absorbing at 442 nm, as attributed to pyridine N-oxides.

The N-oxide of the tertiary aliphatic amine orphenadrine, on the other hand, was devoid of complexing activity, while both orphenadrine per se and the analogous N-demethylated secondary N-hydroxylamine were active complex formers [3].

G. Compounds Containing the N-N Functional Group

There exist reports of cytochrome P450 MI complex formation during the metabolism of hydrazines, N-nitroso compounds and nitrosoureas (Appendix, table I.F). With the hydrazines, complex formation requires O_2 and NADPH reduction of the cytochrome [38], while complex formation from the N-nitrosamines requires NADPH reduction but occurs under anaerobic as well as aerobic conditions [1].

It should be noted that complexes form only from asymmetrical hydrazines. Thus, the presence of the free terminal amino group is essential for the reaction to occur. The formation of complexes from hydrazines has recently [18] also been established in vivo.

It was reported on the basis of in vitro experiments that the binding of N-nitrosamines to cytochrome P450 is followed by reactions which ultimately lead to a denaturation of the enzyme [102].

H. Nitrogen Heterocycles

Many nitrogen heterocycles are found among the compounds which act as ligands to the oxidized (Fe^{3+}) heme of cytochrome P450. Well-known examples are metyrapone [37, 49] and the ellipticines [60, 66]. These type II ligands, however, also have a pronounced affinity for the reduced (Fe^{2+}) form of the heme (Appendix, table I.G). Upon reduction of the oxidized enzyme-substrate complex with NADPH or dithionite, complexes of considerable strength and stability, with absorbance maxima ranging from 440 to 446 nm, are formed.

I. Tertiary Amine-Containing Macrolide Antibiotics

The discovery by a group of French investigators [8, 15, 17, 72, 84 to 87] showing that several macrolide antibiotics (Appendix, table I.H) are important progenitors of cytochrome P450 MI complex formation greatly increased interest in this type of interaction. Since these compounds are widely used as therapeutic agents, their transformation into metabolites inactivating the enzymes may elicit severe drug interactions.

A common characteristic of all the macrolides generating the MI complexes is the presence of an *N,N*-dimethylamino function in one of the sugar residues. NADPH-dependent oxidative transformation of the dimethylamino function is considered to be a prerequisite for the complex formation with the enzyme.

V. Structure-Activity Relationships

When discussing structure-activity relationships (SAR) with respect to cytochrome P450 MI complex formation, it must be realized that this is a rather complicated matter. Although some complex formation occurs in hepatic microsomes from most animal species, variations in substrate efficacy between preparations are frequently noted [30]. Among the complicating factors are the various forms of the cytochrome P450 enzymes that exhibit differences not only in ligand binding but also in the generation of the ultimate ligands from precursor substrates. Accordingly, comparisons of substrates undergoing this reaction have to be restricted to identical enzyme preparations.

Most studies directed at an understanding of the structural requirements for MI complex formation have so far been restricted to aliphatic amines, in a few cases together with their N-oxidized congeners. The most detailed studies have been devoted to amphetamine-type compounds. In addition, SAR have to some extent been implied in the MI complex formation by SKF 525-A and related compounds and with orphenadrine and its congeners. The subject is discussed in reviews by Franklin [30] and Testa and Jenner [102].

A. Amphetamine Class Compounds

One of the most striking features in SAR of complex formation with this class of compounds is that removal of the α-methyl group of amphetamine (2-phenylethylamine), as well as inclusion of an additional α-methyl group (phentermine), prevents the formation of complexes [30]. Substitution of the α-methyl group for larger α-alkyl substituents generally increases the rate of the reaction [64]. The presence of a hydroxyl group in the aromatic ring or on the β-carbon prevents complex formation from the amines, while methoxy substitution in the ring again increases the reaction [30, 102]. The presence of halogen atoms in the aromatic ring (Br, Cl) or on the β-carbon atom (F), although not preventing the complex formation, affects the rate at which the ligand is generated [30].

When relationships between kinetic constants and some physicochemical properties of a homologous series of α-alkyl substituted 2-phenylethylamines were sought, a correlation between the octanol-buffer distribution coefficient (K_D) and rate ($V_{max(obs)}$) of complex formation was evident [64] (fig. 7). When in the same study the corresponding N-hydroxylamines were compared, complex formation was a much less selective phenomenon and the rates, without exception, were quite high. The results suggest that the relative lipophilicity of the substrates is a determining factor in the generation of MI complexes from the amines. Low lipophilicity could then be one explanation why 2-phenylethylamine, *p*-hydroxyamphetamine and norephedrine are inactive as MI complex-forming substrates and why amphetamine shows a low activity.

Although MI complexes are formed also from secondary and tertiary amphetamines, it is difficult to interpret SAR with respect to the degree of nitrogen substitution. It was believed in the early studies on benzphetamine [106] that N-dealkylation was a prerequisite for complex formation from the tertiary amines. However, when using benzphetamine in certain enzyme preparations, there was not a clear correlation occurred between N-dealkylation and complex formation. Thus, complex formation increased with substrate concentrations up to 50 μ*M*, while higher substrate concentrations suppressed the generation of the

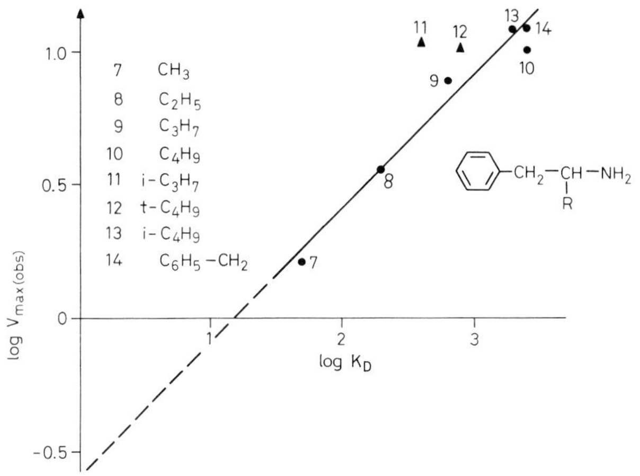

Fig. 7. Correlation between the 1-octanol/buffer partition properties (log K_D) and rate of cytochrome P450 MI complex formation (log $V_{max(obs)}$) for the racemic compounds 7–14.

ligand. The results of more recent studies [46, 47] have shed some light on these early findings and suggest that benzphetamine and norbenzphetamine act by a dual mechanism. The former is, in certain enzyme preparations, a substrate inhibitor of the oxidized form of certain cytochrome P450 enzymes (cf. fig. 3), while the latter prevents benzphetamine metabolism through MI complex formation. Accordingly, the kinetics of complex formation become somewhat complicated when the parent amine and its primary metabolite – both interrelated precursors to the actual ligand – compete for a single enzyme.

B. Other Aliphatic Amines

A common property of aliphatic amines, other than the 1-benzylalkylamines, capable of forming MI complexes is that they are without exception rather lipophilic. The presence of an ethylamino group as a central function, as in SKF 525-A (15) [30] and orphenadrine (16) [3] (fig. 8), has been considered essential [30]. Also, 4,4-diphenyl-2-butylamine (17) generates the complexes [Lindeke, unpublished results]. In general, the rate of MI complex formation is greater for secondary than for either tertiary or primary amines [102]. However, for benzphetamine and norbenzphetamine (section V.A), the interrelationship between tertiary and secondary amines in generating complexes is

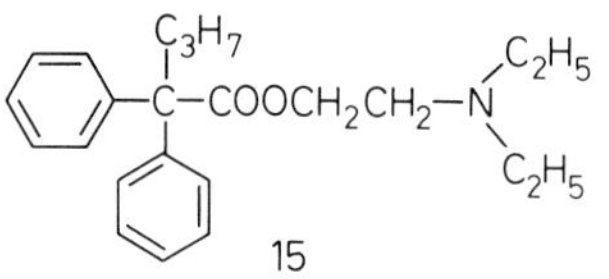

15

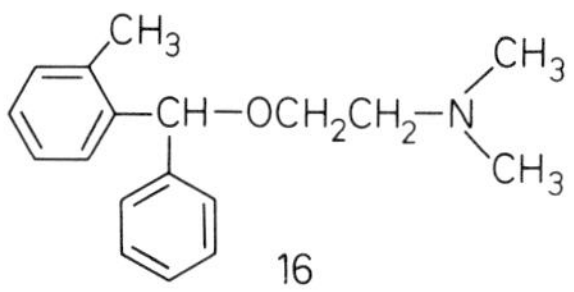

16

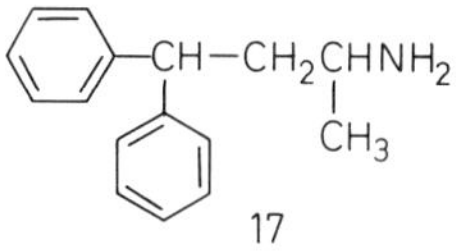

17

Fig. 8. The structures of the aliphatic amines SKF 525-A (15), orphenadrine (16) and 4,4-diphenyl-2-butylamine (17).

complicated by both a substrate and product concentration dependency for a number of amine pairs. For example, tofenacine – the N-demethylated analogue of orphenadrine (16) – exhibited in rat liver microsomes a maximum in the rate of complex formation at a substrate concentration of 33 μM, while normal saturation kinetics occurred with the parent compound [3, 4].

In a later study [5] with orphenadrine analogues with various N-alkyl substituents, a parabolic relationship between log percent cytochrome P450 complexed and hydrophobic fragmental constants was noted. The results from the study do indicate that steric factors beside lipid solubility govern the complexing activity.

Few N-hydroxylamines, apart from those within the amphetamine class of compounds, have been prepared and investigated with respect to complex formation. N-Hydroxytofenacine, a secondary N-hydroxylamine [3], was, however, shown to be orders of magnitude more active than the corresponding secondary amine.

C. Hydrazines

Due to the sequence of events leading to MI complex formation from hydrazines (see section VII.C), only 1-substituted hydrazines are compatible with the suggested mechanism. Within this class of compounds the most rapid complex formation is observed with the N-amino derivatives of saturated N-heterocyclic compounds. Nine hydrazine derivatives were investigated under identical experimental conditions [38] (table I). The results show that the hydrophilic hydrazines were less active and that the activity is also governed by steric factors as indicated

Table I. Relative ability of some 1,1-disubstituted hydrazines to form cytochrome P450 MI complexes in liver microsomes from phenobarbital-pretreated rats

Hydrazine[1]	Relative extent of complex formation
N-Aminopiperidine	1
N-Aminomorpholine	0.14
N-Aminopyrrolidine	0.91
1-Amino-4-methylpiperidine	0.90
1-Amino-2,6-dimethylpiperidine	0.44
1-Amino-4-(β-hydroxyethyl)piperazine	0.22
1-Methyl-1-phenylhydrazine	0.19
1,1-Dimethylhydrazine	0.23

[1] All substrate concentrations were 1 m*M*. Constructed from data reported by Hines and Prough [38].

by the relative activity of N-aminopiperidine versus 1-amino-2,6-dimethylpiperidine.

D. Enantioselectivity

Most studies on MI complex formation using amines with asymmetric centers have so far been performed with racemic substrate mixtures. Thus, there are limited data available on the stereochemical preferences and the only substrates investigated in this respect are again the amphetamine class of compounds.

Enantioselectivity in complex formation has been seen with *R* and *S* benzphetamine and *R* and *S* amphetamine, but with the ratio being dependent on the source of the microsomes [30]. Studies on the influence of the chirality of amphetamine in rabbit liver microsomes have shown that the *R*(−) enantiomer is a better substrate than the *S*(+) enantiomer [25, 44]. In liver microsomes from phenobarbital-pretreated rats, where MI complex formation from amphetamine is considerably less abundant, the differences in rate of complex formation between enantiomers are small, the *S*(+) enantiomer being the more active [25]. However, by increasing the lipophilicity of the substrate, more pronounced differences in rates of formation are observed for the enantiomers in microsomal preparations from this species. Thus, investigations [81] of the optical isomers of the 1-benzylalkylamines shown in figure 9 revealed that the

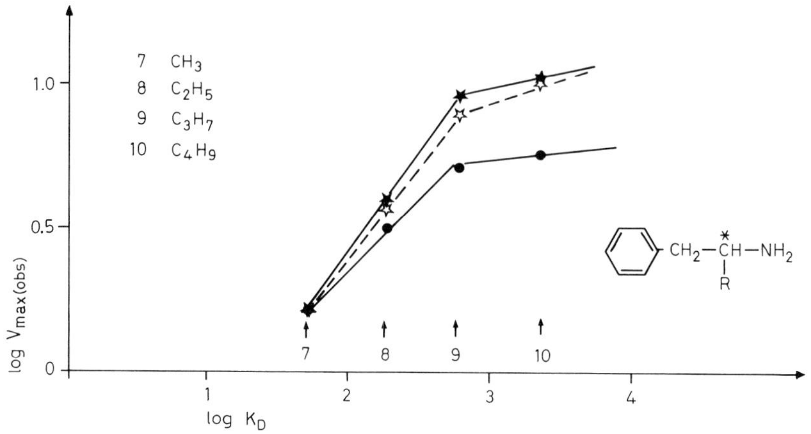

Fig. 9. Correlation between the 1-octanol/buffer partition properties (log K_D) and rate of cytochrome P450 MI complex formation (log $V_{max(obs)}$) for compounds *R* (●) and *S* (★) 7–10 and their racemates (☆) (cf. fig. 7).

S(+) enantiomers produce the MI complexes at a higher rate than the *R*(−) enantiomers, the ratio increasing with increasing size of the α-alkyl substituent. The rates of complex formation for the faster *S*(+) enantiomers coincide with those of the corresponding racemates.

Studies with another lipophilic amphetamine analogue, 1-(2,5-dimethoxy-4-methylphenyl)-2-aminopropane (18) (fig. 10), in rabbit liver microsomes [75] showed that the initial rate of complex formation from the *R*(−) enantiomer was five times as fast as that of the *S*(+) isomer. In this case, the rate of complex formation by the racemic mixture was shown to follow that of the *S*(+) isomer, indicating that complex formation from the *R*(−) enantiomer was inhibited by the *S*(+) enantiomer.

Complex formation by the primary amphetamines appears to follow the species differences seen in deamination and N-oxidation reactions [24, 55, 75]. However, there is evidence [23] that part of the MI complexes formed from N-hydroxyamphetamine in rabbit liver microsomes involves a specific pool of cytochrome P450 enzymes which catalyze N-demethylation of benzphetamine but not the C- or N-oxidation of amphetamine. Moreover, there appears to be no correlation between MI complex formation and disappearance of substrate amines in the metabolism of the racemates of 7–14 (cf. fig. 7) in liver microsomes from phenobarbital-pretreated rats [64].

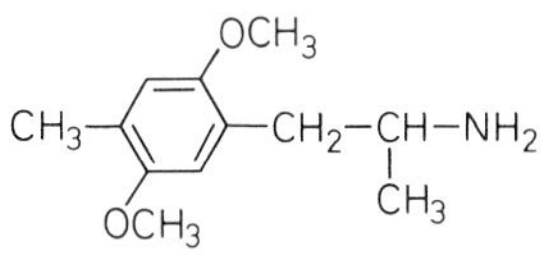

18

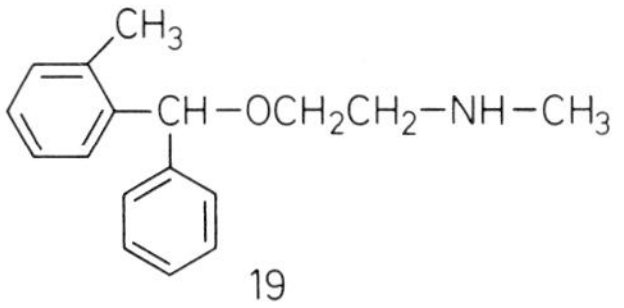

19

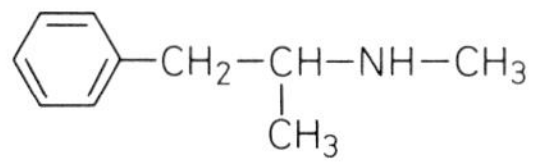

20

Fig. 10. The structures of the aliphatic amines 1-(2,5-dimethoxy-4-methylphenyl)-2-aminopropane (18), tofenacine (19) and methamphetamine (20).

Enantioselectivity in the MI complex formation from the N-hydroxylamines hitherto investigated – N-hydroxyamphetamine [51] and the N-hydroxy analogue of 18 [75] – is not obvious. For both compounds the two enantiomers appear to generate complexes at similar rates and to similar extent. Since complex formation appears sensitive to the chirality of the amines but less so to that of the N-hydroxylamines, the enantioselective formation of the complexes from the amines has been suggested to reflect only the N-hydroxylation step [51, 75] occurring prior to the metabolic reaction generating the ligand. However, the stereochemical requirements for these reactions are more complicated. Thus, other studies [81, 82] have revealed that the *S* enantiomer of 2-nitroso-1-phenylpropane, the C-nitroso analogue of *S*(+) amphetamine, is considerably more active than the *R* enantiomer as a ligand to cytochrome P450 present in liver microsomes from phenobarbital-pretreated rats. In a reconstituted system containing the purified phenobarbital-inducible isozyme, the *S* enantiomer was active while the *R* enantiomer was completely *inactive.*

VI. The Heterogeneity of Ligand-Binding Hemoproteins and Ligand Generation

There are, not unexpectedly, pronounced differences in the interactions of nitrogenous compounds of different oxidation states with hemoproteins. Thus, some classes of nitrogenous compounds interact with a variety of hemoproteins while others generate ligand complexes only with the cytochrome P450 enzymes and eventually only defined subpopulations of the latter.

Aliphatic nitro compounds which were shown to generate complexes with cytochrome P450 enzymes in reducing conditions [68], also form analogue complexes with hemoglobin and myoglobin under such conditions [69]. Similarly, the direct interaction between N-phenylhydroxylamine and cytochrome P450 in the oxidized (Fe^{3+}) state, which results in complex formation, has its precedence in the well-known [56] interaction of this hydroxylamine with methemoglobin. This to some extent is true also for nitrosobenzene [33]. These compounds then behave similarly to NO or CO, both of which have a high affinity for many hemoporphyrins (cf. section III). N-Nitrosoamines [2] and nitrosoureas [90] form ligand complexes in their interactions with both reduced cytochrome P450 and with hemoglobin. Methylhydrazine (cf. section IV.G) reacts with hemoglobin and myoglobin in their Fe^{3+} or Fe^{2+} states, but the reaction requires limited amounts of oxygen [73].

In contrast, complex formation from N-hydroxyamphetamine have been reported to occur only with the cytochrome P450 enzymes, and in rabbit liver microsomes apparently involves only certain specific isozymes [23]. Moreover, the stereochemical preferences seen in the ligand binding of 2-nitroso-1-phenylpropane in liver microsomes from phenobarbital-pretreated rats [81, 82] indicate subtle differences in the interaction of this substrate with the isozymes of cytochrome P450.

It was recognized [26, 30] that different enzyme-inducing agents affect MI complex formation during metabolism of various amines. Thus, phenobarbital pretreatment of experimental animals produces in many cases [3, 26, 46, 83, 92, 108] cytochrome P450 isozymes with a greater ability to generate the complexes than the constitutive enzymes. Other inducers [72, 83, 84, 109] are also active. That both the rate and extent of complex formation by an amine will differ with the source of the cytochrome P450 [8] should thus be viewed as a rule rather than an exception. This is reflected also in the noted species variations [30].

The heterogeneity in cytochrome P450 MI complex formation could reflect differences in the metabolic conversion of an original substrate to the actual ligand, rather than in the ligand-heme interaction itself. Even if it is difficult to assess the contribution of direct and indirect mechanisms, one would expect to see a greater degree of heterogeneity for those substrates which undergo multistep reactions and for which optional metabolic routes not leading to MI complexes are possible.

VII. Events Leading to Ligand Formation

The fact that certain nitrogenous compounds can function as direct ligands to porphyrin enzymes, while others have to undergo metabolic transformation before they form cytochrome P450 ligands (cf. sections II, IV), inevitably raises the question of the interrelationships between structures and the structural determinants for ligand activity. Early studies of MI complex formation considered for the most part various types of aliphatic amines. Franklin [30] originally adopted the MI complex term for those substrates which required oxidative metabolism for activation – i.e. the presence of cytochrome P450, molecular oxygen and NADPH. However, nitro compounds (cf. section IV.E) are also substrates for reactions leading to ligand formation but only under reducing conditions. Thus, with a broader definition for the MI complex term, they can be included among the MI complex-forming substrates.

Amines and nitro compounds represent nitrogenous compounds where the carbon-bound nitrogen attains its lowest and its highest oxidation state [64] (chapters I and II in this volume). Based on the notion that both of these classes of nitrogenous compounds generate cytochrome P450 ligands but after oxidation and reduction, respectively, it was then suggested [51, 67] that the ultimate ligands formed from primary amines and nitro compounds could actually be identical, with the nitrogen atom residing at an intermediary oxidation state.

A. C-N Compounds Containing a Primary Nitrogen Atom

The consecutive stepwise one-electron oxidation of primary amines follows the sequence in equation 2:

$$\underset{(19)}{R{-}\ddot{N}H_2} \Rightarrow \underset{(20)}{R{-}\dot{N}H} \Rightarrow \underset{(21)}{R{-}\ddot{N}HOH} \Rightarrow \underset{(22)}{R{-}\dot{N}OH} \Rightarrow \underset{(23)}{R{-}\ddot{N}{=}O} \Rightarrow \underset{(24)}{R{-}\dot{N}\bar{O}_2} \Rightarrow \underset{(25)}{R{-}NO_2} \quad (2)$$

Studies [74] with model complexes have shown that nitroso alkanes (23) constitute a class of ligands with a very high affinity for Fe^{2+}-porphyrins. The nitroso compounds per se are isoelectronic with dioxygen. The C-nitroso compounds act as ligands to the reduced iron of cytochrome P450, hemoglobin and myoglobin [68, 69, 80] and could hence be regarded as directly acting ligands with reduced cytochrome P450.

These properties of the nitroso alkanes, the oxidation requirement for amines as well as N-hydroxylamines, and the reduction requirement for nitro compounds are consistent with the notion that C-nitroso compounds are the ligands responsible for the MI complexes obtained from primary aliphatic amines or their N-hydroxylamines [51, 62, 68, 69, 71]. However, although there is little doubt that N-oxidation is a prerequisite for MI complex formation from primary amines, the formation of ligands other than the C-nitroso compounds can as yet not be excluded. Other possible ligands are the nitroxide radical (22, equation 2) [51] or intermediates where alterations also at the α-carbon next to the nitrogen have occurred.

B. C-N Compounds Containing a Tertiary or Secondary Nitrogen Atom

The sequence of events leading to ligands for MI complex formation in the metabolism of tertiary and secondary amines is somewhat more obscure than for the primary amines. Werringloer and Estabrook [106, 108] concluded from early studies with benzphetamine that the substrate underwent at least two consecutive metabolic transformations prior to MI complex formation and that initial N-demethylation was essential.

Subsequent studies [3, 63] have shown that there is a correlation between the N-demethylation and concomitant MI complex formation for the two secondary amines tofenacine (19) and methamphetamine (20) (fig. 10). This would suggest that the primary amines are intermediates in reactions leading to the MI complex formation from tertiary or secondary amines, but this suggestion is repudiated by available experimental results [3, 39, 46, 47]. Direct N-oxidation seems to be essential also in this case.

The consecutive oxidation at nitrogen in a secondary, or an N-dealkylated tertiary, amine (cf. chapter I) follows the sequence in equation 3:

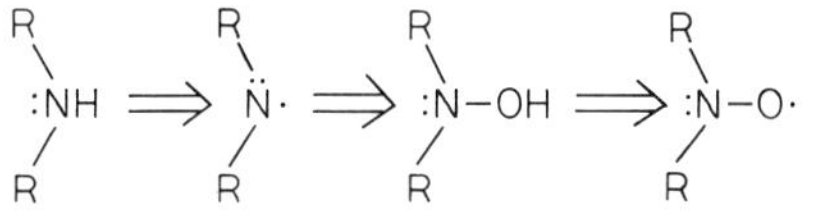

(3)

(26) (27) (28) (29)

If α-hydrogens are available, the nitroxide radical (29) will then be further converted to a nitrone (30, equation 4).

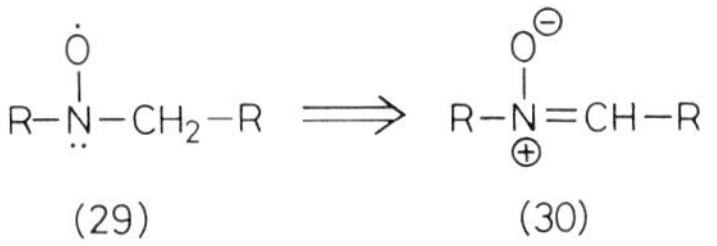

(4)

Secondary N-hydroxylamines as well as nitrones appear as distinct intermediates in the microsomal oxidation of secondary amines [61]. The occurrence of nitrones was first indicated by Kadlubar et al. [53] who showed that both N-benzyl-N-methylhydroxylamine and benzhydryl-N-methylhydroxylamine underwent oxidation.

The formation of nitrones has also been proposed in the microsomal oxidation of N-benzylamphetamine [6, 7, 39, 46, 47, 91] and methamphetamine [63, 91]. Of the two possible nitrones that can be formed in the oxidation of N-hydroxymethamphetamine (4, fig. 6), only nitrone 5b appears to be of metabolic importance. Subsequent hydrolysis of the nitrone to N-hydroxyamphetamine should then be compatible with the noted correlation between MI complex formation and N dealkylation of the secondary amines. If this sequence of events is applicable one would obtain the same precursor for the MI-complex-forming reaction from the tertiary, secondary and primary amines. In fact, subsequent studies [46, 47] have indicated that direct oxidative conversion of the nitrone (31) to the ultimated ligand – believed to be the primary nitroso compound (32) (equation 5) – is more likely in the MI complex formation from benzphetamine.

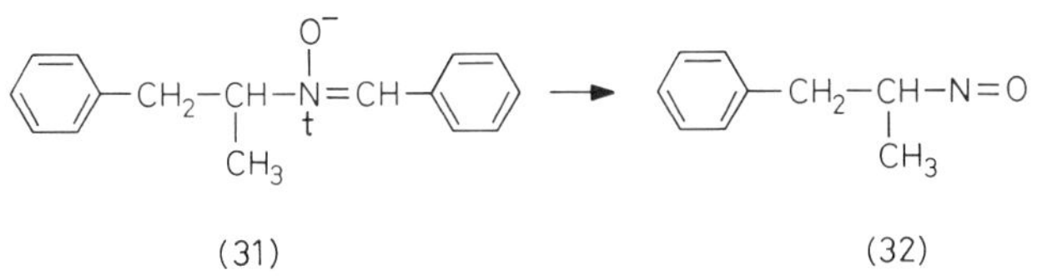

(5)

Similar events could apply to MI complex formation from the vast majority of tertiary and secondary amines, including the tertiary amine-containing macrolide antibiotics [88].

C. N-N Compounds Containing a Primary Nitrogen Atom

That 1,1- but not 1,2-disubstituted hydrazines generate cytochrome P450 MI complexes is in accordance with the mechanism that is believed to be involved in the formation of the ligand. According to this mechanism [38], 1,1-disubstituted hydrazines undergo a two-electron oxidation

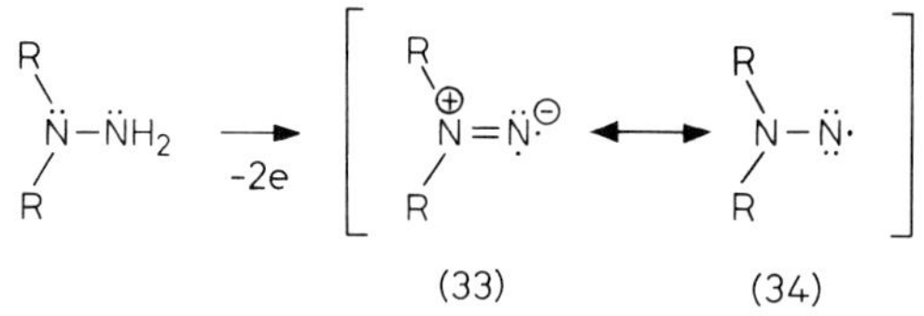

(6)

generating a diazene (34) or a nitrene (33) with high affinity for the heme (equation 6) (cf. also chapter VI in this volume).

However, nitrogen-centered free radicals have also been encountered in the oxidative metabolism of, for example, phenylhydrazine [36, 59, 98] and hydralazine [98] although in noncytochrome P450 systems. As was the case with the nitroxide radical (see section VII. A), there is not enough evidence to exclude the involvement of such radicals.

Cytochrome P450 MI complex formation in the metabolism of N-nitrosamines is consistent with concomitant denitrosation [2] and formation of nitric oxide which is a directly acting ligand to various porphyrin enzymes [20, 21, 34, 48, 78].

VIII. Conclusion

While the interactions between different hemoproteins and ligands, such as CN^-, NO and isocyanides, have been recognized and understood for quite some time, the basis for the spectral perturbations associated with the interaction of various amines with cytochrome P450 during metabolism were not well understood 10 years ago. The knowledge which has accumulated since then indicates, however, that ligands similar to those known earlier to interact with the porphyrins are formed in reactions preceding MI complex formation. Once the ligand is formed, one fundamental principle should be its incorporation as the sixth ligand in a hexacoordinated porphyrin system.

As further insight is gained, cytochrome P450 MI complex formation must be expected as a common event in the metabolism of numerous xenobiotics, particularly those containing nitrogenous functions. That nitrogenous compounds should be especially active as a ligand is a result of the chemistry of organic nitrogen.

Substrates from which MI complexes are formed should be regarded as progenitors of strong enzyme inhibitors, if not suicide substrates, inasmuch as the enzyme is sequestered in the complex. However, when several metabolic steps precede the formation of the ultimate ligand and when some of these are catalyzed by *different isozymes,* only the cytochrome catalyzing the final step is likely to be sequestered. It should be noted that the formation of N-hydroxylamines, i.e. conversion of 19 to 21 in equation 2 and of 26 to 28 in equation 3, and the formation of nitroso compounds and nitrones, i.e. conversion of 21 to 23 in equation 2 and of 28 to 30 in equations 3 and 4 are chemically distinct. The first case is an oxygenation, where oxygen is inserted on the nitrogen atom and the second case is a dehydrogenation. N-Oxidation is catalyzed by at least two classes of microsomal enzymes, the flavin-dependent amine oxidases (chapter IV) and cytochrome P450 (chapter II). Both can participate in pre-MI complex-forming reactions, but by definition only the cytochromes can form complexes.

Our present knowledge of MI complex formation has given a new dimension to the concept of drug interactions [31, 83, 109] but its toxicological consequences have yet to be examined. Many unanswered questions remain. What determines the subpopulations of cytochrome P450 enzymes that form MI complexes? Which factors govern the structural limitation in the ligand with regard to steric hindrance, polarity and stereochemistry? Hopefully, results from experiments with purified enzyme systems and substrates that permit ligand binding in situ will broaden our insight.

Appendix to Table I. Ability of Nitrogenous Compounds to Form Cytochrome P450 Complexes.

Table I.A. Aliphatic amines

Substrate	λ_{max}[a] nm	Lag time s	Ref.
Primary amines			
1. The amphetamine group			
R,S-Amphetamine			25
R-Amphetamine		100	25
S-Amphetamine		60[b]	25
R,S-[2^2H]-Amphetamine		50[b]	51

Table I.A. (cont.)

Substrate	λ_{max}[a] nm	Lag time s	Ref.
1.1 Aromatic substitution pattern			
p-Chloro-amphetamine		slight	25
p-Hydroxy-amphetamine	inactive		25
p-Methoxy-amphetamine			30
m,p-Methylenedioxy-amphetamine		slight	25
Various di- or tri-methylenedioxy-amphetamines			30
R,S o-Methoxy-*m*-methyl-amphetamine		138[c]	75
R o-Methoxy-*m*-methyl-amphetamine		14[c]	75
S o-Methoxy-*m*-methyl-amphetamine		138[c]	75
o- (or *p*-)Hydroxy-*p*- (or-*o*-)-methoxy-amphetamine	inactive		75
o-p-Dihydroxy-amphetamine	inactive		75
p-Carboxy-*o*-,*m*-dimethoxy-amphetamine	inactive		75
1.2 Aliphatic substitution pattern			
2-Phenylethylamine	inactive		25
Tyramine	inactive		25
Dopamine	inactive		25
p-Hydroxy-*m*-methoxy-2-phenylethylamine	inactive		25
Various 1-alkylsubstituted 2-phenylethylamines		30–120[d]	64
Phentermine	inactive		51
2-Fluoro-(or 2,2-difluoro)-2-phenylethylamine			30
2,2-Difluoro-*p*-hydroxy-2-phenylethylamine			30
Norephedrine	inactive		25
2-Hydroxy-*p*-hydroxy-(*m*-hydroxy or *m*-methoxy)-2-phenylethylamine	inactive		25
2. The SKF-525 A group			
DPEA	inactive		28
SKF 26754-A			96
3. Miscellaneous			
bisNororphenadrine	slight		3,4
n-Octylamine	427, 445		45
4,4-Diphenyl-2-butylamine			21
Secondary amines			
1. The amphetamine group			
R,S-Methamphetamine		120–140	25
S-Methamphetamine			44
1.1 Aromatic substitution pattern			25
p-Hydroxy-methamphetamine	inactive		25
p-Chloro-methamphetamine		170	25
o-Methoxy-methamphetamine		180	

Table I.A. (cont.)

Substrate	λ_{max}[a] nm	Lag time s	Ref.
1.2 Aliphatic substitution pattern			
Mephentermine	inactive		25
Ephedrine	inactive		25
m-Hydroxy-2-hydroxymethamphetamine	inactive		25
1.3 Nitrogen substitution pattern			
N-Propylamphetamine			30
R,S-Norbenzphetamine		20	25
S-Norbenzphetamine			44
m-Trifluoromethyl-*N*-ethylamphetamine		50	25
Phenmetrazine			30
1.4 Cyclopentanamine		150	25
Propylhexedrine		160	25
2. The SKF 525-A group			
Dichloro-6-phenylphenoxy-ethylethylamine			28
Fluoxetine			30
Noracetylmethadol			30
Norpropoxyphene			30
SKF 8742-A			96
3. The imipramine group			
Desipramine			12
Nortryptyline			12
4. Miscellaneous			
Tofenacine and analogues			3–5
Tertiary amines			
1. The amphetamine group			
R,S-Benzphetamine		65	106
S-Benzphetamine			44
2. The SKF 525-A group			
Acetylmethadol			30
Adiphenine			12
Benactyzine			12
Cyclizine	inactive		12
Diphenhydramine	slight	~100	5,12
Isomethadone	inactive		12
Lilly 18947		50–60	28
Methadone	slight		12

Table I.A. (cont.)

Substrate	λ_{max}[a] nm	Lag time s	Ref.
2. The SKF 525-A group (cont.)			
Propoxyphene			12
SKF 525-A	454	70–135	96
Trimethobenzamide	inactive		12
3. The imipramine group			
Amitriptyline	inactive		12
Chlorprothixene	inactive		12
Imipramine	inactive		12
Promazine	inactive		12
Promethazine	inactive		12
Trimeprazine	inactive		12
4. Miscellaneous			
Orphenadrine		~100	3–5

[a] Between 455 and 457 nm if not otherwise stated.
[b] Rat.
[c] Rabbit.
[d] Length of the lag time depending on substrate lipophilicity.

Table I.B. Aromatic amines

Substrate	λ_{max}[a] nm	Ref.
Primary amines		
Aniline	(425)445	43
p-Chloro-aniline	448	29
p-Chloro-*o*-methyl-aniline	448	29
Dapsone	450	29
Sulfanilamide	452	29
Secondary amines		
N-Methylaniline	448	41
2-Methylindoline	450–451	29
Tertiary amines		
N,N-Dimethylaniline	448	41

[a] Between 455 and 457 nm if not otherwise stated.

Table I.C. N-Hydroxylamines

Substrate	λ_{max}[a] nm	Ref.
Aliphatic N-hydroxylamines		
1. The amphetamine group		
R,S-N-Hydroxyamphetamine		27
R- and *S*-N-Hydroxyamphetamine		51
1.1 Aromatic substitution pattern		
R,S-*o*-Methoxy-*m*-methyl-N-hydroxyamphetamine		75
R- and *S*-*o*-Methoxy-*m*-methyl-N-hydroxyamphetamine		75
p-Methoxy-N-hydroxyamphetamine		108
1.2 Aliphatic substitution pattern		
N-Hydroxy-2-phenylethylamine		51
Various 1-alkyl-substituted N-hydroxy-2-phenylethylamines		64
N-Hydroxynorephedrine		64
N-Hydroxyphentermine	inactive	51
1.3 Nitrogen substitution		
N-Hydroxy-methamphetamine		61
N-Hydroxy-norbenzphetamine		39
N-Hydroxytofenacine		3–5
1.4 Miscellaneous		
N-*t*-Butylhydroxylamine	inactive	71
N-Methylhydroxylamine	453	63, 71
N-Propylhydroxylamine		68, 71
Aromatic N-hydroxylamines		
N-Phenylhydroxylamine	454	70
p-Chloro-N-phenylhydroxylamine	454	70

[a] Between 455 and 457 nm if not otherwise stated.

Table I.D. C-Nitroso compounds and oximes

Substrate	λ_{max}[a] nm	Ref.
1. C-Nitroso compounds		
2-Nitroso-1-phenylpropane		51, 80
Nitroso-benzene	454	29
p-Chloro-nitroso-benzene	454	70
2. Oximes		
Phenylacetoneoxime	inactive	68
3. Miscellaneous		
Nitric oxide	443	20
Nitrite	444	54

[a] Between 455 and 457 nm if not otherwise stated.

Table I.E. Nitrones, nitro compounds and N-oxides

Substrate	λ_{max}[a] nm	Ref.
1. Nitrones		
N-Benzylidene-1-phenyl-2-propylamine N-oxide		39
N-Methylene-1-phenyl-2-propylamine N-oxide		63
N-(1-Phenyl-2-propylidene)-methylamine N-oxide	453	63
2. Aliphatic nitro compounds		
2-Methyl-2-nitro-3-phenylpropane	inactive	67
2-Methyl-2-nitro-propane	inactive	67
Nitro-amphetamine		67
Nitro-cyclohexane	422, 461	67, 93
Nitro-cyclopentane		67
Nitro-ethane		67
Nitro-*n*-hexane		67
Nitro-methane		67
Nitro-*n*-pentane		67
2-Nitro-1-phenylethane		68
2-Nitro-propane	454	104
Nitro-toluene		67
Phenyl-nitromethane	423, 451	93
Tetranitromethane	424, 454	93

Table I.E. (cont.)

Substrate	λ_{max}[a] , nm	Ref.
3. Aromatic nitro compounds		
Nitro-benzene	400 (454)	33, 100
p-Chloro-nitro-benzene	400 (454)	70
p-Nitro-benzoic acid	400	100
o-Nitro-fluorene	400	100
o-Nitro-naphtalene	400	100
4. N-Oxides		
N,N-Dimethylaniline N-oxide	448	41
Orphenadrine N-oxide	inactive	3, 4
Pyridine N-oxide	442	42
Tiaramide N-oxide	442	99

[a] Between 455 and 457 nm if not otherwise stated.

Table I.F. Compounds containing the *N-N*-functional group

Substrate	λ_{max}[a] nm	Ref.
1. 1,1-Disubstituted hydrazines		
N-Amino-homopiperidine	438, 449	38
l-Amino-4- (β-hydroxyethyl)-piperazine	438, 449	38
l-Amino-4-methylpiperidine	438, 449	38
l-Amino-2,6-dimethylpiperidine	438, 449	38
N-Amino-morpholine	438, 449	38
N-Amino-piperidine	438, 449	38
N-Amino-pyrrolidine	438, 449	38
N-Methyl-N-formylhydrazine	425, 447	10
1,1-Dimethyl-hydrazine	438, 449	38
1-Methyl-1-phenylhydrazine	438, 449	38
2. 1,2-Disubstituted hydrazines		
1,2-Dimethylhydrazine	inactive	38
Procarbazine	inactive	38

Table I.F. (cont.)

Substrate	λ_{max}[a] nm	Ref.
3. Monosubstituted hydrazines		
Benzyl-hydrazine	inactive	38
Butyl-hydrazine	inactive	38
Ethyl-hydrazine	inactive	38
Methyl-hydrazine	weak to inactive	18, 38
Phenyl-hydrazine	444	18, 50, 105
4. N-Nitroso compounds		
Dibutyl-nitrosamine	444–450	1
Diethyl-nitrosamine	450–453	94
Dimethyl-nitrosamine	450–453	94
Diphenyl-nitrosamine	443	1
Dipropyl-nitrosamine	444–450	1
Nitrosomorpholine	447	1
5. Miscellaneous		
1-(2-Chloroethyl)-3-cyclohexyl-1-nitrosourea	444	90
1,3-bis(2-Chloroethyl)-1-nitrosourea	444	90

[a] Between 455 and 457 nm if not otherwise stated.

Table I.G. Nitrogen heterocycles

Substrate	λ_{max}[a] nm	Ref.
Ellipticine	445	60
9-Methoxy (or 9-hydroxy)-ellipticine	445	60
Metyrapone	446	37
Nicotinamide	446[b]	49
Pyridine	446[b]	43, 49
3-(3-Pyridyl)-propanol	446[b]	49
1-(2-Pyridyl)-2-(3-or 4-pyridyl)-ethylene	446[b]	49
Pyrazinamide	442	22

[a] Between 455 and 457 nm if not otherwise stated.
[b] The maximal band obtained at pH 7.8 in microsomes from phenobarbital treated rats.

Table I.H. Macrolide antibiotics

Substrate	λ_{max}[a] nm	Lag time[b] s	Ref.
Tertiary amines			
Erythromycin			15
Josamycin	inactive		58
Lincomycin	inactive		16
Methylenecycline	inactive		16
Methymycin			16
Midecamycin	inactive		58
Oleandamycin			87
Rifampicin	inactive		16
Spiramycin	inactive		16
Tetracycline	inactive		16
Troleandomycin			17, 85, 109
Tylosin	inactive		16

[a] Between 455 and 457 nm if not otherwise stated.
[b] Self-induction.

References

1 Appel, K.E.; Ruf, H.H.; Mahr, B.; Schwarz, M.; Rickart, R.; Kunz, W.: Binding of nitrosamines to cytochrome P-450 of liver microsomes. Chem.-Biol. Interact. *28:* 17–33 (1979).

2 Appel, K.E.; Graf, H.: Metabolic nitrite formation from N-nitrosamines: evidence for a cytochrome P-450 dependent reaction. Carcinogenesis *3:* 293–296 (1982).

3 Bast, A.; Noordhoek, J.: Spectral interaction of orphenadrine and its metabolites with oxidized and reduced hepatic microsomal cytochrome P-450 in the rat. Biochem. Pharmacol. *31:* 2745–2753 (1982).

4 Bast, A.; Van Kemenade, F.A.A.; Savenije-Chapel, E.M.; Noordhoek, J.: Product inhibition in orphenadrine metabolism as a result of a stable cytochrome P-450-metabolic intermediary complex formed during the disposition of mono-N-desmethyl-orphenadrine (Tofenacine) in the rat. Res. Commun. chem. Pathol. Pharmacol. *40:* 391–403 (1983).

5 Bast, A.; Savenije-Chapel, E.M.; Nordhoek, J.: Relationship between molecular structure and cytochrome P-450-metabolic intermediary complex formation, studied with orphenadrine analogues. J. pharm. Sci. *73:* 953–956 (1984).

6 Beckett, A.H.; Gibson, G.G.: Identification of the in vitro N-oxidized metabolites of (+)- and (−)-N-benzylamphetamine. J. Pharm. Pharmacol. *29:* 756–760 (1977).

7 Beckett, A.H.; Gibson, G.G.: Stereochemistry of the metabolic incorporation of oxygen into (+)- and (−)-N-benzylamphetamine. Xenobiotica *8:* 73–83 (1978).

8 Bonfils, C.; Dalet-Beluche, I.; Maurel, P.: Induction by triacetyloleandomycin and partial purification of a LM3 form of cytochrome P-450 from rabbit liver microsomes. Biochem. biophys. Res. Commun. *104:* 1011–1017 (1982).

9 Boogerd, F.C.; Verseveld, H.W. van; Stouthamer, A.H.: Electron transport to nitrous oxide in *Paracoccus denitrificans.* FEBS Lett. *113:* 279–284 (1980).

10 Braun, R.; Greff, U.; Netter, K.J.: Indications for nitrosamide formation from the mushroom poison gyromitrin by rat liver microsomes. Xenobiotica *10:* 557–574 (1980).

11 Brown, S.B.: Ultraviolet and visible spectroscopy; in Brown, An introduction to spectroscopy for biochemists, pp. 14–69 (Academic Press, London 1980).

12 Buening, M.K.; Franklin, M.R.: SKF 525-A inhibition, introduction, and 452-nm complex formation. Drug Metab. Disposit. *4:* 244–255 (1976).

13 Dahl, A.R.; Hodgson, E.: Complexes of stannous fluoride and other group IVB dihalides with mammalian hemoproteins. Science *197:* 1376–1378 (1977).

14 Dahl, A.R.; Hodgson, E.: Complexes of trivalent oxygenated phosphorus compounds with cytochrome P-450 and cytochrome P-420: the origin of double soret spectra. Chem.-Biol. Interact. *21:* 137–154 (1978).

15 Danan, G.; Descatoire, V.; Pessayre, D.: Self-induction by erythromycin of its own transformation into a metabolite forming an inactive complex with reduced cytochrome P-450. J. Pharmac. exp. Ther. *218:* 509–514 (1981).

16 Delaforge, M.; Jaouen, M.; Mansuy, D.: Dual effects of macrolide antibiotics on rat liver cytochrome P-450. Induction and formation of metabolite-complexes: a structure-activity relationship. Biochem. Pharmacol. *32:* 2309–2318 (1983).

17 Delaforge, M.; Jaouen, M.; Mansuy, D.: The cytochrome P-450 metabolite complex derived from troleandomycin: properties in vitro and stability in vivo. Chem.-Biol. Interact. *51:* 371–376 (1984).

18 Delaforge, M.; Battioni, P.; Mahy, J.P.; Mansuy, D.: In vivo formation of σ-methyl and σ-phenyl-ferric complexes of hemoglobin and liver-cytochrome P-450 upon treatment of rats with methyl- and phenylhydrazine. Chem.-Biol. Interact. *60:* 101–114 (1986).

19 Donovan, J.W.: Ultraviolet difference spectroscopy – new techniques and applications; in Hirs, Timasheff, Methods in enzymology, vol. XXVII, pp. 497–525 (Academic Press, London, 1973).

20 Ebel, R.E.; O'Keeffe, D.H.; Peterson, J.A.: Nitric oxide complexes of cytochrome P-450. FEBS Lett. *55:* 198–201 (1975).

21 Ehrenberg, A.; Szczepkowski, T.W.: Properties and structure of the compounds formed between cytochrome *c* and nitric oxide. Acta chem. scand. *14:* 1684–1692 (1960).

22 Facino, R.M.; Carini, M.: The inhibitory effect of pyrazinamide on microsomal monooxygenase activities is related to the binding to reduced cytochrome P-450. Pharmacol. Res. Commun. *12:* 523–537 (1980).

23 Florence, V.M.; Cho, A.K.; Effect of 455 nm complex on amphetamine metabolism by rabbit liver microsomes. Fed. Proc. *41:* 1730 (1982).

24 Florence, V.M.; Di Stefano, E.W.; Sum, C.Y.; Cho, A.K.: The metabolism of (R)-(−)-amphetamine by rabbit liver microsomes. Initial products. Drug Metab. Disposit. *10:* 312–315 (1982).

25 Franklin, M.R.: Complexes of metabolites of amphetamines with hepatic cytochrome P-450. Xenobiotica *4:* 133–142 (1974).

26 Franklin, M.R.: The influence of cytochrome P-450 induction on the metabolic formation of 455-nm complexes from amphetamines. Drug Metab. Disposit *2:* 321–326 (1974).
27 Franklin, M.R.: The formation of a 455 nm complex during cytochrome P-450-dependent N-Hydroxyamphetamine metabolism. Mol. Pharmacol. *10:* 975–985 (1974).
28 Franklin, M.R.: Inhibition of the metabolism of N-substituted amphetamines by SKF 525-A and related compounds. Xenobiotica *4:* 143–150 (1974).
29 Franklin, M.R.: A new class of inhibitory cytochrome P-450 complexes formed during metabolism: a comparison with amphetamine and SKF 525-A type complexes. Chem.-Biol. Interact. *14:* 337–346 (1976).
30 Franklin, M.R.: Inhibition of mixed-function oxidations by substrates forming reduced cytochrome P-450 metabolic-intermediate complexes. Pharmacol. Ther. A *2:* 227–245 (1977).
31 Franklin, M.R.: Ligand complexes generated during cytochrome P-450 dependent metabolism; in Snyder, Parke, Kocsis, Jollow, Gibson, Witmer, Biological reactive intermediates-II, part A, pp. 165–177 (Plenum Publishing, New York 1982).
32 Fry, I.V.; Cammack, R.; Hucklesby, D.P.; Hewitt, E.J.: Stability of the nitrosyl-sirohaem complex of plant nitrite reductase, investigated by EPR spectroscopy. FEBS Lett. *111:* 377–380 (1980).
33 Fukuto, J.M.; Brady, J.F.; Burstyn, J.N.; Van Atta, R.B.; Valentine, J.S.; Cho, A.K.: Direct formation of complexes between cytochrome P-450 and nitrosoarenes. Biochemistry *25:* 2714–2719 (1986).
34 Garrioch, A.G.; Hewitt, E.J.: Annu. Rep. Long Ashton Res. Stn, pp. 58–59 (1976).
35 Gumbrecht, J.R.; Franklin, M.R.: The formation of cytochrome P-450-metabolic intermediate complexes from amines, in the isolated perfused rat liver. Xenobiotica *9:* 547–554 (1979).
36 Hill, H.A.; Thornalley, P.J.: Phenyl radical production during the oxidation of phenylhydrazine and in phenylhydrazine-induced haemolysis. FEBS Lett. *125:* 235–238 (1981).
37 Hildebrandt, A.G.; Leibman, K.C.; Estabrook, R.W.: Metyrapone interaction with hepatic microsomal cytochrome P-450 from rats treated with phenobarbital. Biochem. biophys. Res. Commun. *37:* 477–485 (1969).
38 Hines, R.N.; Prough, R.A.: The characterization of an inhibitory complex formed with cytochrome P-450 and a metabolite of 1,1-disubstituted hydrazines. J. Pharmac. exp. Ther. *214:* 80–86 (1980).
39 Hirata, M.; Lindeke, B.; Orrenius, S.: Cytochrome P-450 product complexes and glutathione consumption produced in isolated hepatocytes by norbenzphetamine and its N-oxidized congeners. Biochem. Pharmacol. *28:* 479–484 (1979).
40 Hlavica, P.: Biological oxidation of nitrogen in organic compounds and disposition of N-oxidized products. CRC crit. Rev. Biochem. *12:* 39–101 (1982).
41 Hlavica, P.; Aichinger, G.: Studies on the cytochrome P-450 product complexes formed during the metabolism of *N,N*-dimethylaniline. Biochim. biophys. Acta *544:* 185–199 (1978).
42 Hlavica, P.; Mietaschk, J.; Baden, I.: Interaction of ligands with cytochrome P-450. On the 442 nm spectral species generated during the oxidative metabolism of pyridine. Biochem. J. *204:* 425–432 (1982).

43 Imai, Y.; Sato, R.: Anomalous spectral interactions of reduced P-450 with ethyl isocyanide and some other lipophilic ligands. J. Biochem., Tokyo *62:* 464–473 (1967).

44 James, R.C.; Franklin, M.R.: Comparisons of the formation of cytochrome P-450 complexes absorbing at 455 nm in rabbit and rat microsomes. Biochem. Pharmacol. *24:* 835–838 (1975).

45 Jefcoate, C.R.E.; Gaylor, J.L.; Calabrese, R.L.: Ligand interactions with cytochrome P-450. I. Binding of primary amines. Biochemistry *8:* 3455–3463 (1969).

46 Jeffery, E.H.; Mannering, G.J.: Sharing of cytochrome P-450 isozymes during the stepwise microsomal oxidation of benzphetamine; in Hietanen, Laitanen, Hanninen, Cytochrome P-450, biochemistry, biophysics and environmental applications, pp. 59–62 (Elsevier/North-Holland, Amsterdam 1982).

47 Jeffery, E.H.; Mannering, G.J.: Interaction of constitutive and phenobarbital-induced cytochrome P-450 isozymes during the sequential oxidation of benzphetamine: explanation for the difference in benzphetamine-induced hydrogen peroxide production and 455-nm complex formation in microsomes from untreated and phenobarbital-treated rats. Mol. Pharmacol. *23:* 748–757 (1983).

48 John, M.E.; Waterman, R.: Structural characteristics of nitrosyl hemoglobins and their relation to ESR spectra. FEBS Lett. *106:* 219–222 (1979).

49 Jonen, H.G.; Huthwohl, B.; Kahl, R.; Kahl, G.F.: Influence of pyridine and some pyridine derivates on spectral properties of reduced microsomes and on microsomal drug metabolizing activity. Biochem. Pharmacol. *23:* 1319–1329 (1974).

50 Jonen, H.G.; Werringloer, J.; Prough, R.A.; Estabrook, R.W.: The reaction of phenylhydrazine with microsomal cytochrome P-450. Catalysis of heme modification. J. biol. Chem. *257:* 4404–4411 (1982).

51 Jonsson, J.; Lindeke, B.: On the formation of cytochrome P-450 product complexes during the metabolism of phenylalkylamines. Acta pharm. suec. *13:* 313–320 (1976).

52 Jung, C.; Friedrich, J.; Ristau, O.: Quantum chemical interpretation of the spectral properties of the CO and O_2 complexes of hemoglobin and cytochrome P-450. Acta biol. med. germ. *38:* 363–377 (1979).

53 Kadlubar, F.F.; McKee, E.M.; Ziegler, D.M.: Reduced pyridine nucleotide-dependent N-hydroxy amine oxidase and reductase activities of hepatic microsomes. Archs Biochem. Biophys. *156:* 46–57 (1973).

54 Kahl, R.; Wulff, U.; Netter, K.J.: Effect of nitrite on microsomal cytochrome P-450. Xenobiotica *8:* 359–364 (1978).

55 Kammerer, R.C.; Jonsson, J.; Gal, J.; Cho, A.K.: Current concepts: use of stable isotopes in studies on the metabolism of amphetamine. Life Sci. *23:* 283–290 (1978).

56 Keilin, D.; Hartree, E.F.: Reactions of hemoglobin and its derivatives with phenyl hydroxylamine and nitrosobenzene. Nature *151:* 390–391 (1943).

57 Kumaki, K.; Nebert, D.W.: Spectral evidence for weak ligand in sixth position of hepatic microsomal cytochrome P-450 low spin ferric iron in vivo. Pharmacology *17:* 262–279 (1978).

58 Larrey, D.; Tinel, M.; Pessayre D.: Formation of inactive cytochrome P-450 Fe(II)-metabolite complexes with several erythromycin derivatives but not with josamycin and midecamycin in rats. Biochem. Pharmacol. *32:* 1487–1493 (1983).

59 Lebedeva, O.V.; Ugarova, N.N.; Berezin, I.V.: Irreversible inactivation of heme-containing peroxidase in the presence of phenylhydrazine and the protective effect of free radical reaction inhibitors. Role of heme-phenylhydrazine complex formation in heme degradation. Biokhimiya *44:* 2235–2244 (1979).
60 Lesca, P.; Lecointe, P.; Paoletti, C.; Mansuy, D.: Ellipticines as potent inhibitors of aryl hydrocarbon hydroxylase: their binding to microsomal cytochrome P-450 and protective effects against benzo(a)pyrene mutagenicity. Biochem. Pharmacol. *27:* 1203–1209 (1978).
61 Lindeke, B.: The non- and postenzymatic chemistry of N-oxygenated molecules. Drug Metab. Rev. *13:* 71–121 (1982).
62 Lindeke, B.; Jonsson, J.; Hallström, G.; Paulsen, U.: The effect of α-carbon substitution on the 455 nm complexes formed by phenethylamines and their N-oxidized congeners; in Gorrod, Biological oxidation of nitrogen, pp. 47–51 (Elsevier/North-Holland, Amsterdam 1978).
63 Lindeke, B.; Paulsen, U.; Anderson, E.: Cytochrome P-455 complex formation in the metabolism of phenylalkylamines. IV. Spectral evidences for metabolic conversion of methamphetamine to N-hydroxyamphetamine. Biochem. Pharmacol. *28:* 3629–3635 (1979).
64 Lindeke, B.; Paulsen-Sörman, U.; Hallström, G.; Khuthier, A.-H.; Cho, A.K.; Kammerer, R.C.: Cytochrome P-455-nm complex formation in the metabolism of phenylalkylamines. Drug Metab. Disposit. *10:* 700–705 (1982).
65 Mannering, G.J.: Microsomal enzyme systems which catalyze drug metabolism; in LaDu, Mandel, Way, Fundamentals of drug metabolism and drug disposition, pp. 206–252 (William & Wilkins, Baltimore 1971).
66 Mansuy, D.: Coordination chemistry of cytochrome P-450 and iron-porphyrins: relevance to pharmacology and toxicology. Biochemie *60:* 969–977 (1978).
67 Mansuy, D.; Beaune, P.; Chottard, J.-C.; Bartoli, J.-F.; Gans, P.: The nature of the '455 nm absorbing complex' formed during the cytochrome P-450 dependent oxidative metabolism of amphetamine. Biochem. Pharmacol. *25:* 609–612 (1976).
68 Mansuy, D.; Gans, P.; Chottard, J.-C.; Bartoli, J.-F.: Nitrosoalkanes as Fe(II) ligands in the 455-nm-absorbing cytochrome P-450 complexes formed from nitroalkanes in reducing conditions. Eur. J. Biochem. *76:* 607–615 (1977).
69 Mansuy, D.; Chottard, J.C.; Chottard, G.: Nitrosoalkanes as Fe(II) ligands in the hemoglobin and myoglobin complexes formed from nitroalkanes in reducing conditions. Eur. J. Biochem. *76:* 617–623 (1977).
70 Mansuy, D.; Beaune, P.; Cresteil, T.; Bacot, C.; Chottard, J.-C.; Gans, P.: Formation of complexes between microsomal cytochrome P-450-Fe-(II) and nitrosoarenes obtained by oxidation of arylhydroxylamines or reduction of nitroarenes in situ. Eur. J. Biochem. *86:* 573–579 (1978).
71 Mansuy, D.; Rour, R.; Bacot, C.; Gans, P.; Chottard, J.-C.; Leroux, J.-P.: Interaction of aliphatic N-hydroxylamines with microsomal cytochrome P-450: nature of the different derived complexes and inhibitory effects on monooxygenase activities. Biochem. Pharmacol. *27:* 1229–1237 (1978).
72 Mansuy, D.; Delaforge, M.; LeProvost, E.; Flinois, J.P.; Columelli, S.; Beaune, P.: Induction of cytochrome P-450 in rat liver by the antibiotic troleandomycin: partial purification and properties of cytochrome P-450 troleandomycin metabolite complexes. Biochem. biophys. Res. Commun. *103:* 1201–1208 (1981).

73 Mansuy; D.; Battioni, P.; Mahy, J.-P.; Gillet, G.: Comparison of the hemoglobin reactions with methyl- and phenylhydrazine: intermediate formation of a hemoglobin Fe(II)-methyldiazene complex. Biochem. biophys. Res. Commun. *106:* 30–36 (1982).

74 Mansuy, D.; Battioni, P.; Chottard, J.-C.; Riche, C.; Chiaroni, A.: Nitrosoalkane complexes of iron-porphyrins: analogy between the bonding properties of nitrosoalkanes and dioxygen. J. Am. chem. Soc. *105:* 455–463 (1983).

75 McGraw, N.P.; Castagnoli, N., Jr.: Studies on chiral interactions of 1-(2,5-dimethoxy-4-methylphenyl)-2-aminopropane and the corresponding N-hydroxy metabolites with cytochrome P-450. J. med. Chem. *24:* 299–304 (1981).

76 Omura, T.; Sato, R.: The carbon monoxide-binding pigment of liver microsomes. II. Solubilization, purification, and properties. J. biol. Chem. *239:* 2379–2385 (1964).

77 Orrenius, S.; Ernster, L.: Microsomal cytochrome P-450 linked monooxygenase systems in mammalian tissues; in Hayaishi, Molecular mechanisms of oxygen activation, pp.. 215–244 (Academic Press, New York 1974).

78 Paul, P.; Kumta, U.S.: Reduced binding of nitric oxide in irradiated horse heart myoglobin. J. agric. Fd Chem. *23:* 37–40 (1975).

79 Pauling, L.: The nature of the chemical bond; 3rd ed., pp. 145–182 (Cornell University Press, Ithaca 1960).

80 Paulsen-Sörman, U.; Lundkvist, G.; Khuthier, A.-H.; Lindeke, B.: Cytochrome P-455 nm complex formation in the metabolism of phenylalkylamines. VII. Enzyme interactions with synthetic 2-nitroso-1-phenylalkanes. Chem.-biol. Interact. *47:* 1–14 (1983).

81 Paulsen-Sörman, U.; Jönsson, K.-H.; Lindeke, B.: Cytochrome P-455-nm complex formation in the metabolism of phenylalkylamines. VIII. On the stereoselectivity in metabolic intermediary complex formation with a series of chiral 2-substituted 1-phenylaminoethanes. J. med. Chem. *3:* 342–346 (1984).

82 Paulsen-Sörman, U.; Jönsson, K.-H.; Lindeke, B.: Cytochrome P-455 nm complex formation in the metabolism of phenylalkylamines IX. A comparative study with enantiomeric 2-nitroso-1-phenylpropane in microsomes and a reconstituted cytochrome P-450 system from rat liver. Chem. biol. Interact. (in press).

83 Pershing, L.K.; Franklin, M.R.: Cytochrome P-450 metabolic-intermediate complex formation and induction by macrolide antibiotics; a new class of agents. Xenobiotica *12:* 687–699 (1982).

84 Pessayre, D.; Descatoire, V.; Konstantinova-Mitcheva, M.; Wandscheer, J.-C.; Cobert, B.; Level, R.; Benhamou, J.-P.; Jaouen, M.; Mansuy, D.: Self-induction by triacetyloleandomycin of its own transformation into a metabolite forming a stable 456-nm-absorbing complex with cytochrome P-450. Biochem. Pharmacol. *30:* 553–558 (1981).

85 Pessayre, D.; Konstantinova-Mitcheva, M.; Descatoire, V.; Cobert, B.; Wandscheer, J.-C.; Level, R.; Feldmann, G.; Mansuy, D.; Benhamou, J.-P.: Hypoactivity of cytochrome P-450 after triacetyloleandomycin administration. Biochem. Pharmacol. *30:* 559–564 (1981).

86 Pessayre, D.; Larrey, D.; Vitaux, J.; Breil, P.; Belghiti, J.; Benhamou, J.-P.: Formation of an inactive cytochrome P-450 Fe(II)-metabolite complex after administration of troleandomycin in humans. Biochem. Pharmacol. *31:* 1699–1704 (1982).

87 Pessayre, D.; Descatoire, V.; Tinel, M.; Larrey, D.: Self-induction by oleandomycin of its own transformation into a metabolite forming an inactive complex with reduced cytochrome P-450. Comparison with troleandomycin. J. Pharmac. exp. Ther. *221:* 215–221 (1982).
88 Pessayre, D.; Tinel, M.; Larrey, D.; Cobert, B.; Funkbretano, C.; Babany, G.: Inactivation of cytochrome P-450 by a troleandomycin metabolite. Protective role of glutathione. J. Pharmac. exp. Ther. *224:* 685–691 (1983).
89 Peterson, J.A.: The ligands of cytochrome P-450: identification and role in the activation of dioxygen; in Caughey, Biochemical and clinical aspects of oxygen, pp. 227–262 (Academic Press, New York 1979).
90 Potter, D.W.; Reed, D.J.: Denitrosation of carcinostatic nitrosoureas by purified NADPH cytochrome P-450 reductase and rat liver microsomes to yield nitric oxide under anaerobic conditions. Archs. Biochem. Biophys. *216:* 158–169 (1982).
91 Prough, R.A.; Ziegler, D.M.: The relative participation of liver microsomal amine oxidase and cytochrome P-450 in N-demethylation reactions. Archs Biochem. Biophys. *180:* 363–373 (1977).
92 Roberts, S.M.; Franklin, M.R.: Modification of hepatic microsomal oxidative drug metabolism in rats by the opiate maintenance drugs acetylmethadol, propoxyphene, and methadone. Life Sci. *25:* 845–852 (1979).
93 Sakurai, H.; Hermann, G.; Ruf, H.H.; Ullrich, V.: The interaction of aliphatic nitro compounds with liver microsomal monooxygenase system. Biochem. Pharmacol. *29:* 341–345 (1980).
94 Saprin, A.N.; Ramseyer, J.; McConn, J.; Piette, L.H.: Electron spin resonance and optical spectra of the complexes of dialkylnitrosoamines with cytochrome P-450. Biochem. biophys. Res. Commun. *77:* 789–796 (1977).
95 Schenkman, J.B.; Remmer, H.; Estabrook, R.W.: Spectral studies of drug interaction with hepatic microsomal cytochrome. Mol. Pharmacol. *3:* 113–123 (1967).
96 Schenkman, J.B.; Wilson, B.J.; Cinti, D.L.: Diethylaminoethyl 2,2-diphenylvalerate HCl (SKF 525-A) – in vivo and in vitro effects of metabolism by rat liver microsomes – formation of an oxygenated complex. Biochem. Pharmacol. *21:* 2373–2383 (1972).
97 Schenkman, J.B.; Sligar, S.G.; Cinti, D.L.: Substrate interaction with cytochrome P-450. Pharmacol. Ther. *12:* 43–71 (1981).
98 Sinha, B.K.; Motten, A.G.: Oxidative metabolism of hydralazine. Evidence for nitrogen centered radical formation. Biochem. biophys. Res. Commun. *105:* 1044–1051 (1982).
99 Sugiura, M.; Iwasaki, K.; Noguchi, H.; Kato, R.: Evidence for the involvement of cytochrome P-450 in tiaramide N-oxide reduction. Life Sci. *15:* 1433–1442 (1974).
100 Sternson, L.A.; Gammans, R.E.: Interaction of aromatic nitro compounds with reduced hepatic microsomal cytochrome P-450. Drug Metab. Disposit. *3:* 266–274 (1975).
101 Swinyard, E.A.: Noxious gases and vapours. Carbon monoxide, hydrocyanic acid, gasoline, kerosene, carbon tetrachloride, and miscellaneous organic solvents; in Goodman, Gillman, The pharmacological basis of therapeutics; 4th ed., pp. 930–943 (Macmillan, London 1970).
102 Testa, B.; Jenner, P.: Inhibitors of cytochrome P-450s and their mechanisms of action. Drug Metab. Rev. *12:* 1–117 (1981).

103 Ullrich, V.: The mechanism of cytochrome P-450 action; in Ullrich, Roots, Hildebrandt, Estabrook, Conney, Microsomes and drug oxidations, pp. 192–201 (Pergamon Press, Oxford 1977).
104 Ullrich, V.; Schnabel, K.H.: Formation and bonding of carbanions by cytochrome P-450 of liver microsomes. Drug Metab. Disposit. *1:* 176–183 (1973).
105 Ullrich, V.; Hermann, G.; Weber, P.: Nitrite formation from 2-nitropropane by microsomal monooxygenases. Biochem. Pharmacol. *27:* 2301–2304 (1978).
106 Werringloer, J.; Estabrook, R.W.: Evidence for an inhibitory product cytochrome P-450 complex generated during benzphetamine metabolism by liver microsomes. Life Sci. *13:* 1319–1330 (1973).
107 Werringloer, J.; Estabrook, R.W.: Heterogeneity of liver microsomal cytochrome P-450: the spectral characterization of reactants with reduced cytochrome P-450. Archs Biochem. Biophys. *167:* 270–286 (1975).
108 Werringloer, J.; Estabrook, R.W.: The characterization of product adducts of liver microsomal cytochrome P-450 and their use as probes for the heterogeneity of cytochrome P-450 as modified by the induction of drug metabolism; in Estabrook, Lindenlaub, The induction of drug metabolism, pp. 269–307 (Schattauer Verlag, Stuttgart 1978).
109 Wrighton, S.A., Maurel, P.; Schuets, E.G.; Watkins, P.B.; Young, B.; Guzellah, P.S.: Identification of the cytochrome P-450 induced macrolide antibiotics in rat liver as the glucocorticoid responsive cytochrome P-450p. Biochemistry *24:* 2171–2178 (1985).

B. Lindeke, PhD, Medical Department, Aco Lakemedel AB,
Box 3026, S-171 03 Solna (Sweden)

Prog. bas. clin. Pharmac., vol. 1, pp. 103–131 (Karger, Basel 1988)

IV. Nitrogen Oxygenation by Flavin Enzymes

R.A. Prough

Department of Biochemistry, The University of Louisville School of Medicine, Louisville, Ky., USA

Introduction

Historical

With the advent of the modern consumer society, thousands of new compounds have been synthesized in the quest for improved chemical intermediates, cosmetics, drugs, and food additives. Many of these chemicals contain the heteroatom, nitrogen, and much concern exists related to the carcinogenic and toxic potential of these nitrogenous compounds. The purpose of this review is to describe a unique flavin-dependent enzyme which is involved in a large number of N- and *S*-oxygenation[1] reactions. While some of the reactions catalyzed by flavin-dependent oxidases may lead to deleterious results in vivo, many N-oxygenated products have chemical properties which allow them to be efficiently and rapidly excreted.

The existence of biological N-oxygenation reactions was postulated by Heubner [1913] to account for the active agent(s) of the arylamines which are capable of oxidizing oxyhemoglobin. The arylamines were known to elicit a number of toxic responses as well as induce tumors in animals. Since the pure arylamines themselves were not effective in oxidizing hemoglobins in vitro, Heubner [1913] postulated that an N-oxygenation reaction of arylamines may give hydroxylamines, which would react like phenylhydroxylamine to oxidize hemoglobin. Lintzel [1934] noted that humans excreted large amounts of trimethylamine

[1] The term N-oxygenation refers to reactions in which molecular oxygen is activated to form N-O bonds, as opposed to N-oxidation by removal of electrons from the N center.

N-oxide when given trimethylamine orally. These early studies heralded the finding that enzyme systems exist which can oxygenate compounds containing the heteroatom, nitrogen.

Initial biochemical studies on N-oxygenation arose from diverse areas of interest in the early 1960s. While studying the mechanism of squalene oxidase, Chaykin and Bloch [1959] noted that nicotinamide N-oxide enhanced this enzyme's activity. Subsequently, nicotinamide was noted to yield its N-oxide as an excretory product upon intraperitoneal administration to rats. Baker et al. [1963] continued studies on N-oxide formation, particularly trimethylamine, in plants and animals. Using an ammonium sulfate-precipitated fraction of hog liver, Baker and Chaykin [1962] described an enzyme system which catalyzed trimethylamine N-oxide formation. NADPH and atmospheric oxygen were required for the reaction, indicative of a monooxygenase enzyme mechanism.

During this same period, Kiese and co-workers had endeavored, to define the active principle of mixtures of aniline compounds which could lead to the known toxic reactions of these compounds. Kiese and Uehleke [1961] demonstrated that aniline and N-methylaniline were oxidized to their N-hydroxy derivatives in the presence of liver microsomes, NADPH, and oxygen in a manner similar to trimethylamine. Ziegler and Pettit [1964] first described the N-oxygenation of *N,N*-dimethylaniline to its N-oxide, *N,N*-dimethylaniline N-oxide. This finding was later confirmed by Uehleke et al. [1971]. Since hydroxylamine derivatives are known to interact directly to cause hemoglobin oxidation, these N-oxygenation reactions appeared to form the toxic agents of the aniline derivatives. In addition, Cramer et al. [1960], Weisberger and Weisberger [1973] and Irving [1971] were instrumental in demonstrating that the carcinogenic aromatic amides and amines, such as N-acetylaminofluorene or 4-aminobiphenyl, were N-hydroxylated in an NADPH- and oxygen-dependent microsomal reaction. Bickel [1969] and Kadlubar et al. [1976] showed that these derivatives may be more reactive when the N-hydroxy compounds were further metabolized to sulfate or glucuronide conjugates. Since the initial studies on N-oxygenation reactions involved toxic or carcinogenic compounds, it was assumed by most workers that N-oxygenation reactions would lead to the production of reactive intermediates. However, as pointed out in several reviews by Bickel [1969], Coutts and Beckett [1977] and Ziegler [1980, 1984], many drugs are effectively detoxified and eliminated after they are metabolically converted to the highly polar (and presumedly excretable) N-oxide derivatives. Many of these N-oxides are reasonably nontoxic.

The realization that an NADPH- and O_2-dependent monooxygenase system capable of N-oxygenating organic compounds existed in the microsomal fraction of liver and other tissues merged this research area with the biochemical studies on the cytochrome-P450-dependent enzyme system described by Estabrook et al. [1963] and Omura and Sato [1964]. The plethora of biochemical studies on the cytochrome P450 enzyme system during the 1960s and early 1970s prompted researchers to assume that this hemoprotein-dependent electron transport chain accounted for the N-oxygenation reactions mentioned above and, consequently, underplayed the role of the flavin adenine dinucleotide (FAD)-containing monooxygenase (amine oxidase, N-oxidase, or FAD-containing monooxygenase) described first by Ziegler and Pettit [1966]. The 1970s provided much information related to the biochemical details of the cytochrome-P450-dependent monooxygenase as seen in a recent book by Sato and Omura [1978] and in a major review by White and Coon [1980]. This enzyme system is inducible in vivo by administration of barbiturates, polycyclic aromatic hydrocarbons and other diverse agents, and exists in numerous isozymic forms. In addition, the various forms of cytochrome P450 have been shown to have a very broad substrate specificity. This enzyme system is known to oxygenate endogenous compounds, such as steroids and fatty acids, as well as exogenous compounds, such as carcinogens and drugs.

The possibility that multiple forms of cytochrome P450 exist caused some confusion with early inhibitor studies on N-oxygenation. Both Beckett [1971] and Uehleke [1960, 1971] provided evidence that the cytochrome-P450-dependent monooxygenase participates in the N-hydroxylation of primary arylamines and arylamides, but may not be involved in the N-oxidation of N-alkylanilines and tertiary amines. These results were subsequently encapsulated by Gorrod [1973, 1978a, b] in a theory related to the multiplicity of N-oxygenating enzymes which used as its conceptual basis the differences in pK_a of the nitrogenous compounds. As envisioned by this pK_a concept, the less basic (and less nucleophilic) amines were assumed to be the best substrates for the cytochrome-P450-dependent monooxygenase and the more basic (and nucleophilic) amines were the best substrates for the FAD-containing monooxygenase described by Ziegler and Pettit [1966]. While there are numerous exceptions related to primary alkyl- and arylamines, Gorrod's original hypothesis continues to serve as a general predictive tool. For example, arylamines should be good substrates for FAD-dependent monooxygenase but not for cytochrome P450 based on pK_a. In fact, arylamines are not good substrates for the flavoprotein, but are well metabolized by cytochrome P450 [Ziegler, 1980].

Using immunochemical techniques, Masters and Ziegler [1971] established that the flavoprotein reductase, NADPH-cytochrome *c* (P450) reductase, is not related to the FAD-containing monooxygenase. In hog, they also noted that NADPH-cytochrome *c* reductase, but not N-oxidase, activity was induced by animal pretreatment with phenobarbital. Subsequently, Prough and Ziegler [1977] demonstrated that when one uses a microsomal suspension from rodent liver, both the cytochrome- and FAD-dependent monooxygenases function simultaneously in the metabolism of *sec.*- and *tert.*-amines. Using differential inhibitors of the two monooxygenases, it was shown that the demethylation of benzphetamine catalyzed by cytochrome P450 occurred simultaneous with the formation of the benzphetamine N-oxide catalyzed by the amine oxidase. Similar results have been noted with another tertiary amine, *N,N*-dimethylaniline (see below). The fact that the stable N-oxide is formed simultaneously with *sec.*-amines and formaldehyde suggests that determination of stoichiometry between NADPH or O_2 utilization and formaldehyde formation catalyzed by liver microsomes is not likely to succeed unless the FAD-containing monooxygenase is inhibited. While the specific inhibitory antibody to NADPH-cytochrome *c* (P450) reductase caused better than 85% inhibition of demethylation of *tert.*-amines, such as aminopyrine, benzphetamine, brompheniramine, and ethylmorphine by pig liver microsomes, this antibody inhibited the demethylation of secondary amines by less than 25–50%. This suggested that both monooxygenases catalyze the metabolism of *sec.*-amines, such as *p*-chloro-*N*-methylaniline, ephedrine, desipramine, and N-methylamphetamine. Prough and Ziegler [1977] also demonstrated that the N-hydroxy derivatives of *sec.*-amines formed during metabolism catalyzed by the pure flavoprotein, such as *p*-chloro-*N*-methylaniline, are not stable. In fact, both formaldehyde (15–20%) and N-methylhydroxylamine metabolite (80–85%) can be measured in the water and chloroform phase, respectively, of reaction mixtures extracted with chloroform. However, if strong acids, such as 1% trichloroacetic acid, are added to terminate the reaction, only formaldehyde can be measured due to the acid lability of the N-methylhydroxylamine metabolite.

Table I. N-Oxidation reactions catalyzed by the FAD-containing monooxygenase

Substrate	Product	Reference
1 *tert.*-Amines[a] R_1 \\ N-R_3 / R_2	N-oxides R_1 O \\ + / N / \\ R_2 R_3	Ziegler and Pettit, 1966
2 *sec.*-Amines R_1-NH-R_2	hydroxylamines R_1-N(OH)-R_2	Ziegler, 1980
3 Primary aliphatic hydroxylamines RCH_2-NHOH	oximes RCH=NOH	Ziegler, 1980
4 Hydroxylamines O | R_1-N-CH_2-R_2	nitrones O | R-N=CH-R_2	Poulsen et al., 1974
5 Hydrazines		Prough et al., 1981
A 1,1-Disubstituted R_1 \\ N-NH_2 / R_2 CH_2	diazenium ions (decomposition products) R_1 \\ + N=NH / R_2CH_2 R_1 \\ + N-NH_2 || R_2-CH	
B Mono- and 1,2-disubstituted R_1CH_2-NH-NH-R_4[b]	azo and hydrazone compounds R_1CH_2-N=N-R_4 R_1CH=N-NHR_4	Erickson and Prough, 1986; Prough, 1973; Prough et al., 1981

[a] For most compounds studied, R_1 is CH_3-.
[b] R_4 may be CH_3- or H-.

The studies of Prough and Ziegler [1977] would indicate that in the hog, hamster and rat, little N-oxide formation is due to the cytochrome-P450-dependent monooxygenase. However, the assays reported could not rule out a 5–10% relative participation of cytochrome P450. In fact, Hlavica and Kehl [1977] have demonstrated that partially purified cytochrome P450 from rabbit liver can N-oxidize *N,N*-dimethylaniline, albeit at rates lower than N-demethylation; further consideration of this point will be made elsewhere in this review.

Monoamine Oxidases

The purpose of this review is to address the properties and catalytic reactions of the FAD-containing monooxygenase elegantly described by Ziegler and co-workers. However, one other flavin-containing oxidase must be mentioned since simple dehydrogenation reactions can occur for nitrogen-containing compounds. For example, compounds such as benzylamine [Tabor et al., 1954], monoalkylhydrazines [Erikson and Prough, 1986] or 1,2-disubstituted hydrazines [Coomes and Prough, 1983] can be oxidized to their respective Schiff's base or azo derivatives. These intermediates can subsequently decompose to give aldehydes and the dealkylated nitrogen compounds [Tabor et al., 1954] or can be metabolized further [Wiebkin and Prough, 1980]. Due to the similarity of products of the mitochondrial monoamine oxidase and the microsomal FAD-containing monooxygenase, the mitochondrial oxidase must not be overlooked in amine oxidation.

Flavin-Containing Monooxygenase

The FAD-containing monooxygenase (EC 1.14.13.8) can activate molecular oxygen at the expense of NADPH and subsequently form a coordinate covalent bond between the nitrogen atom and one atom of oxygen (table I). During the last 5 years, several research groups have addressed the chemical and physical properties of this enzyme, but Ziegler and co-workers mastered techniques of purification of this membrane-bound enzyme and described many of its reactions, kinetics, and physical properties. The following review will attempt to explore the chemistry and biochemical mechanism of this unique flavoprotein monooxygenase. Several review articles on the FAD-containing monooxygenase have been published by Ziegler [1978, 1980, 1984] and Ziegler et al. [1971].

The FAD-Containing Monooxygenase

Substrate Specificity

Like the cytochrome P450-dependent enzyme system, one property of the FAD-containing monooxygenase also referred to as the microsomal amine oxidase or N-oxidase, is its relatively broad substrate specificity. Table I encapsulates the major classes of nitrogenous compounds which are catalytically N-oxygenated by this enzyme. Nearly all tertiary amines which do not bear a positive charge can be N-oxygenated, but the affinity of the enzyme for its substrates varies greatly depending on the structure of the substituents. At saturating substrate concentrations, the k_{cat} values (turnover per mol flavin) for nearly all of the reactions catalyzed by the flavoprotein are approximately 34–35 min^{-1} at pH 8.2 and 37 °C. Due to this fact, the specificity of the enzyme for a substrate, often defined as k_{cat}/k_m, is dictated principally by the K_m of the substrate for the enzyme [Ziegler, 1980]. It should be noted in table I that in the case of aliphatic amines, the N-hydroxy products can undergo a second N-oxidation reaction to form either oximes (primary aliphatic hydroxylamines) or nitrones (disubstituted hydroxylamines with at least one aliphatic substituent). Similar results are observed for the N-aminoheterocyclic compounds (hydrazine derivatives such as N-aminopiperidine) [Prough, 1973; Prough et al., 1981] and the sulfur compounds such as thioamides and thiocarbamides [Poulsen et al., 1974a, 1979]. More detailed reviews related to the substrate specificity of the monooxygenase have been published by Ziegler [1980, 1984] and Ziegler et al. [1971]. It is of interest that primary amines, while basic in nature, are not good substrates with the exception of a limited number of arylamines (2-naphthylamine [Ziegler et al., 1973], and aminofluorene [Pelroy and Gandolfi, 1980; Frederick et al., 1982], 2-aminoanthracene [Pelroy and Gandolfi, 1980] and 1-aminopyrene [Pelroy and Gandolfi, 1980]). No stable N-hydroxy or N-oxide intermediates have been demonstrated for the hydrazines, and azo or diazene intermediates most likely are the initial stable products of their N-oxidation [Erikson and Prough, 1986; Prough et al., 1981]. This broad substrate specificity allows the monooxygenase to effectively N-oxygenate a variety of nitrogenous compounds, thereby altering their basicity and lipophilicity.

Localization

Ziegler and Pettit [1966] demonstrated that the FAD-containing monooxygenase is located in the microsomal fraction of liver from pig and rat. From early work on trimethylamine N-oxidase activity [Baker et al., 1963], it was noted that the N-oxygenase activity was associated with the microsomal fraction of many extrahepatic tissues of mammals. Ohmiya and Mehendale [1981] provided evidence that there may be multiple forms of the lung enzyme in rats and rabbits. Recently, Williams et al. [1984] and Tynes et al. [1985] have demonstrated that the enzymes found in mouse or rabbit lung are immunochemically and catalytically distinct. Multiple forms of this flavin monooxygenase may exist between species and organs [Smyser et al., 1985]. The work of Devereux et al. [1985] with lung cells has demonstrated that this FAD-containing monooxygenase is nearly equally distributed between Clara (65%) and type I (35%) alveolar cells, while cytochrome P450 appears to be principally found in Clara cells (90%). These studies were expanded for *N,N*-dimethylaniline N-oxidase activity in a wide variety of mammalian species by Machinist et al. [1968]. Recently, Patton et al. [1980] used fluorescent-labeled antibody to the pure pig liver FAD-containing monooxygenase to determine whether the enzyme is localized in the nucleus. Aside from the high immunofluorescence due to the enzyme on the endoplasmic reticulum, fluorescence intensity could also be shown in the nuclear envelope and may account for up to 10% of the enzyme in hamster hepatocytes.

Purification

The major effort to purify the monooxygenase has been made in Ziegler's laboratory, but successful purification efforts have been reported by Beaty and Ballou [1980] and Prough et al. [1981]. Ziegler and Poulsen [1978] have published a detailed purification procedure. As will be seen later, the enzyme is heat labile and the livers must be removed prior to scalding the hogs to facilitate removal of the hide. This heat lability requires that care be taken in preparation of liver microsomes, especially when certain tissue blenders are used for homogenization. The livers should be cut into small strips. Once chilled, the activity is reasonably stable. The purification requires the use of detergents to extract the monooxygenase and several DEAE-cellulose chromatography steps in the presence of Triton X-100. Beaty and Ballou [1981a] have also used a lauryl-Sepharose-4B column chromatography procedure with success; this procedure removes adventitious cytochrome b_5 which can interfere with spectral studies on the enzyme. Upon purification, the enzyme can be precipitated with 25% polyethylene glycol, resuspended in 0.05 *M* potassium phosphate buffer (pH 7.2), and fractionated on a Bio-Bead SM2 column to remove traces of Triton X-100. The resulting enzyme preparation is stable for several weeks at 2 °C or for months at −20 °C.

Physical Properties

As isolated, the purified mammalian FAD-containing monooxygenase has a minimum molecular weight of approximately 56,000, as measured by polyacrylamide gel electrophoresis after treatment with sodium dodecyl sulfate (SDS) and mercaptoethanol and 64,000 based on the amino acid composition per mole of flavin [Poulsen and Ziegler, 1979]. The flavin content of the enzyme was determined to be approximately 15.2 nmol FAD per mg protein. In the absence of SDS, the enzyme exists as an aggregate which, depending on the protein concentration, can have ranges of molecular weight from $2.4–5.1 \times 10^5$ [Ziegler, personal communication]. The enzyme contains nine sulfhydryl groups which are not available for reaction with thiol reagents until after treatment with SDS. As isolated, the enzyme contains no trace metals which function in the catalytic cycle or as structural elements of the enzyme.

Catalytic Properties

In conjunction with the studies on the physical properties of the FAD-containing monooxygenase, Poulsen and Ziegler [1979] performed a steady-state kinetic analysis of the N-oxygenation of trimethylamine and *S*-oxygenation of methimazole (under conditions where only one *S*-oxygenation step occurred). Based on the initial velocity studies measuring NADPH oxidation at 340 nm, they interpreted their data to indicate that the reaction was an ordered Ter-Bi mechanism with an irreversible step between the addition of the second and third substrate. In this mechanism the oxidized enzyme adds NADPH first, then reacts with molecular oxygen and, finally, adds the substrate to be oxygenated. Since $NADP^+$ was shown to be a competitive inhibitor for NADPH binding, the last product to leave the enzyme must be $NADP^+$, as suggested by the method of Cleland [1967]. Therefore, the N-oxygenated products are presumed to be released from the enzyme prior to release of $NADP^+$. Studies on 1,1-disubstituted hydrazines demonstrated that the kinetic mechanism of hydrazine oxygenation catalyzed by the flavoprotein was identical to that determined with trimethylamine and methimazole [Prough et al., 1981].

In recent years, considerable interest has focused on the mechanism of oxygen activation by flavoprotein monooxygenases. In 1976, a study of the mechanism of *p*-hydroxybenzoate hydroxylase [Entsch et al., 1976], a bacterial FAD-containing monooxygenase, demonstrated the existence of a transient intermediate of the reduced flavoprotein in the presence of oxygen. The intermediate was a derivative of FAD which had an absorbance maxima at 380–390 nm. This derivative could react with *p*-hydroxybenzoate or 2,4-dihydroxybenzoate to give the hydroxylated products, 3,4-dihydroxybenzoate and 2,3,4-trihydroxybenzoate. Ghisla et al. [1977] provided evidence that these intermediates of reduced flavoproteins which absorb light at 380 to 390 nm may well be C(4a)-flavin hydroxides or hydroperoxides. These experiments involved the synthesis of 4a-hydroxy-5-ethyl-3-methyllumiflavin and the pseudobases of the FAD/FMN derivatives and the stoichiometric incorporation of these hydroxy-flavins into the apoproteins of *D*-amino acid oxidase, lactate hydroxylase, and *p*-hydroxybenzoate hydroxylase. Further studies with model compounds [Kemal and Bruice, 1976] and using NMR techniques [Ghisla et al., 1978] have identified the existence of both the C(4a)-hydroperoxyflavin and C(4a)-hydroxyflavin intermediates of luciferase, a bacterial monooxygenase capable of oxidizing aldehydes to acids.

Ball and Bruice [1979, 1980] have investigated the chemical oxidation/oxygenation of amines by C(4a)-flavin hydroperoxides in *t*-butyl alcohol using C(4a)-hydroperoxy-5-ethyl-3-methyllumiflavin (4a-F1Et-OOH). They observed that a wide variety of *sec.*- and *tert.*-amines reacted with the 4a-F1Et-OOH in a bimolecular process to give products identical to those observed by Poulsen et al. [1974b] using the purified monooxygenase. Of interest was the failure of 4a-F1Et-OOH to N-oxygenate pyridine derivatives or primary amines, and the ability of the flavin hydroperoxide to N-oxygenate N-methyl-N-benzylhydroxylamine to form a mixture of nitrones. In addition, the ratios of the α-phenyl-*N*-methylnitrone to α-methyl-*N*-benzylnitrone products for the enzymic [Poulsen et al., 1974b] and chemical [Ball and Bruice, 1980] reactions were 3 and 2.2, respectively. This indicates that the formation of the stable N-oxygenated products of hydroxylamines results from dehydration of a common, enzymatically generated hydroxylamine oxide intermediate. With *N,N*-dimethylaniline and diisopropylamine, Ball and Bruice [1980] showed the stoichiometric formation of the respective N-oxide or hydroxylamine derivative and of C(4a)-hydroxy-5-ethyl-3-methyllumiflavin (4a-F1Et-OH). The chemical mechanism was suggested to involve a nucleophilic displacement of the terminal oxygen of the C(4a)-flavin hydroperoxide by sp^3-hybridized nitrogen (fig. 1). The pK_a hypothesis of Gorrod [1973, 1978a, b] also assumed some correlation between substrate nucleophilicity and reactivity. Doerge and Corbett [1984a, b] have performed a Hammett structure-reactivity study with arylamines and organosulfur compounds. While their results confirm the importance of the nucleophilic attack of nitro-

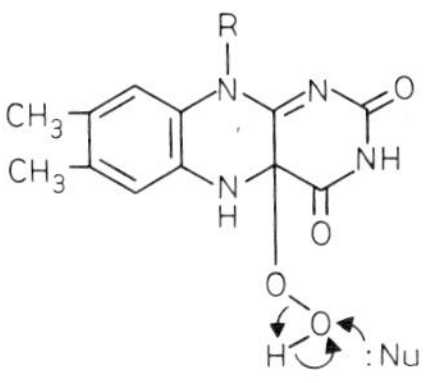

Fig. 1. The scheme of nucleophilic displacement of the terminal oxygen of C(4a)-flavin hydroperoxides by amine substrate as envisioned by Ball and Bruice [1979, 1980]. :Nu = N- or *S*-containing nucleophile.

gen on the 4a-F1Et-OOH, there is evidence that hydrophobic interactions of the substrate with the flavoprotein may modulate the k_{cat}/K_m for a given substrate. These factors would support the preference for polynuclear aromatic amines compared to monoarylamines, such as aniline, or to alkylamines, such as *n*-octylamine.

However, this mechanism of nucleophilic displacement is very different from that postulated for iron porphyrin monooxygenases such as cytochrome P450 [White and Coon, 1980; Shannon and Bruice, 1981]. The reactivity of 4a-F1Et-OOH in such reactions is 10^4–10^5 times greater than that seen for either H_2O_2 or *t*-butyl hydroperoxide [Ball and Bruice, 1979]. The results of this work demonstrate that flavin hydroperoxides are capable of chemically oxygenating amines to form amine oxides or hydroxylamines and have distinct spectral characteristics to aid in evaluating the enzyme mechanism of the microsomal FAD-containing monooxygenase.

Poulsen and Ziegler [1979] and Beaty and Ballou [1980, 1981a] demonstrated the formation of a stable spectral intermediate of the monooxygenase which has an absorbance maxima near 360 nm upon reduction of the enzyme by NADPH and exposure to molecular oxygen (fig. 2). This intermediate was stable in the absence of substrate and was assumed to be the C(4a)-hydroperoxy FAD form of the enzyme. Using spectral and transient-state kinetic techniques, Beaty and Ballou [1980, 1981a, b] have provided considerable detail regarding the reductive and oxidative steps of the enzyme's mechanism. They have shown that the reduction of the flavin is a biphasic process and that the fast phase can be abolished by prior addition of the oxidized pyridine nucleotide, $NADP^+$, to the enzyme. The existence of an oxidized enzyme-$NADP^+$ complex was demonstrated as well. The effect of

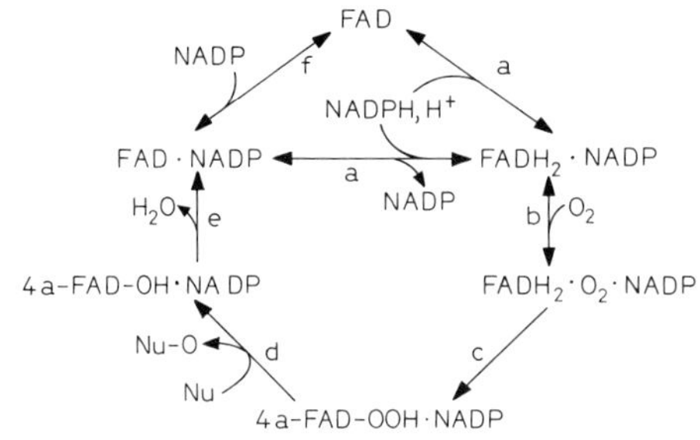

Fig. 2. The catalytic cycle of amine N-oxidation by the microsomal FAD-containing monooxygenase. Step c is assumed to be irreversible and steps b and f are assumed to be reversible [Poulsen and Ziegler, 1979; Beaty and Ballou, 1981a,b]. No direct information is available regarding the reversibiity of steps a, d, and e. 4a-FAD-OOH·NADP is the stable C(4a)-hydroperoxy FAD form of the enzyme and 4a-FAD-OH·NADP is the pseudobase form of the flavoprotein, C(4a)-hydroxy FAD.

$NADP^+$ appears to change the free form of the enzyme in a slow step to an $NADP^+$-bound form which is reduced less well than the free form. In addition, the presence of the substrate to be oxygenated has no effect on the rate of flavin reduction under anaerobic conditions, in agreement with the kinetic mechanism proposed by Poulsen and Ziegler [1979]. Since a primary kinetic isotope effect on the rate of reduction was observed using (4R)-[4-^{2}H]-NADP, the rate-limiting step of reduction may involve the breaking of the C-H bond at the C-4 position of NADPH. An activator of the catalytic N-oxygenation reaction, *n*-octylamine, was shown to cause a 40% decrease in the rate of flavin reduction. However, the rate of flavin reduction is 25 times greater than the catalytic turnover of the enzyme and the inhibitory effect of *n*-octylamine on reduction most likely is not related to the activation seen with this compound (see below).

Since reduced flavin is rapidly oxidized by molecular oxygen to give H_2O_2, the key question to be answered regarding the mechanism of monooxygenases concerns the mode by which the enzyme activates and stabilizes a reactive oxygen species. As shown by Poulson and Ziegler [1979], the enzyme when reduced by NADPH has an NADPH-oxidase activity (in the absence of substrate) which is approximately 5% of the rate of the substrate-dependent reaction. The product of this reaction appears to be largely hydrogen peroxide; however, Rauchman et al. [1979] have shown that superoxide anion accounts for approximately

3.5% of the electrons utilized by the NADP-oxidase activity. In an elegant study, Beaty and Ballou [1981b] have studied the oxidative half-reactions of theFAD-containing monooxygenase. First, the reoxidation of the enzyme in the absence of $NADP^+$ is rapid and not affected by the presence of the substrate to be oxygenated. These results suggest, that, unlike *p*-hydroxybenzoate hydroxylase, the substrate to be oxygenated does not stabilize the activated oxygenated species of the microsomal monooxygenase. However, addition of $NADP^+$ to the reduced enzyme prior to reaction with oxygen allows the formation of a stable intermediate with a spectrum like a C(4a)-hydroperoxy FAD [4a-FAD-OOH] derivative. This hydroperoxy flavin derivative is stabilized by the binding of $NADP^+$ and, in the absence of $NADP^+$, the C(4a)-hydroperoxy FAD spontaneously decays to give hydrogen peroxide [Poulsen and Ziegler, 1979]. As expected, the decomposition of the 4a-FAD-OOH in the presence of $NADP^+$ is largely dependent upon the substrate to be oxygenated, the nucleophile which displaces the terminal oxygen of the stable flavin hydroperoxide.

Poulsen and Ziegler [1979] have shown that certain artificial electron acceptors, particularly dichlorophenolindophenol, can react anaerobically with the reduced enzyme, but react poorly in the presence of oxygen. In agreement, Beaty and Ballou [1981b] have suggested that a reversible binding step of the reduced enzyme with O_2 exists before formation of the stable 4a-FAD-OOH. Prior to addition of oxygen, the $FADH_2$-form of the enzyme can react with exogenous electron acceptors. After binding of oxygen, the reduced flavin is converted to the C(4a)-hydroperoxy FAD derivative which cannot react with exogenous acceptors. The very low rate of superoxide anion formed in the absence of substrates [Rauchman et al., 1979] may be due to the existence of a small steady-state level of the $FADH_2$-O_2 form of the enzyme in equilibrium with the 4a-FAD-OOH adduct.

Finally, Beaty and Ballou [1981b] have noted that during the last step of the reaction the 4a-FAD-OOH form of the enzyme transiently converts (in a lag phase) to another spectral intermediate prior to decomposition to the oxidized flavoprotein. This new spectral intermediate may be a pseudobase, such as a C(4a)-hydroxyflavin or some similar species. The reaction phase following the lag proceeds at a rate which is approximately equal to the steady-state turnover number of the enzyme. This result suggests that the rate-limiting step of the catalytic cycle may be either the rate of substrate oxygenation or the rate of decomposition of the pseudobase, C(4a)-hydroxy FAD. Since the maximal catalytic rate of N-oxygenation by the monooxygenase is fairly constant for all substrates (30–50 nmol product formed per min per mole enzyme), Beaty and Ballou [1981b] suggested that a common event, such as hydrolysis of the pseudobase, might account for the rate-limiting step of the enzyme. While the N-oxygenation step was thought to be less plausible for a common step, a mechanism involving oxygen transfer to the amine substrates as a rate-limiting step cannot be eliminated. Figure 2 provides an overall reaction mechanism to account for the kinetic properties of the enzyme. In vivo, the intracellular concentrations of $NADP^+$ most likely insure that the oxidized form of the enzyme may always have one molecule of $NADP^+$ bound per mole of enzyme, thereby favoring the formation of a stable C(4a)-hydroperoxy flavin form of the enzyme. This situation would insure that no significant amounts of reduced oxygen, such as hydrogen peroxide or superoxide anion, are produced in the absence of a substrate to be oxygenated.

Effectors

Aside from the intriguing role of $NADP^+$ in stabilizing the C(4a)-hydroperoxyflavin form of the enzyme, two other activation phenomena have been noted for the enzyme. Certain nonsubstrate nitrogenous compounds, such as lipophilic primary alkylamines and alkylguanidines, increase the rate of overall N-oxygenation, as well as the k_{cat} [Ziegler et al., 1973, Ziegler and Mitchell, 1972]. Beaty and Ballou [1981b] noted that *n*-octylamine caused a 1.6-fold increase in the rate of reoxidation of the enzyme; i.e. the rate of decomposition of the C(4a)-hydroperoxy-flavin complex (56). Although n-octylamine, the best known effector, inhibits the rate of flavin reduction by 40% and stimulates the decomposition of the C(4a)-hydroperoxyflavin complex 1.6-fold, the effect of *n*-octylamine must be related principally to hydroperoxyflavin decomposition which, when activated, is 8 to 10 times slower than the rate of reduction in the presence of the aliphatic primary amines.

In addition, certain of the poorer substrates for the monooxygenase, such as primary amines and hydrazines [Ziegler and Mitchell, 1972; Poulsen et al., 1976; Prough et al., 1981], not only have larger apparent K_m values, but 1.4. to 1.8-fold larger V_{max} values. These observations suggest that a separate positive effector site may exist in addition to the binding site for $NADP^+$ which stabilizes the C(4a)-hydroperoxyflavin complex.

Method of Assay

Traditional methods for measuring the microsomal monooxygenases have been varied and are limited due to either the formation of multiple products with differing chemical properties, as seen for substrates like the *sec.*-and *tert.*-amines [Prough and Ziegler, 1977], or the poor coupling of the monooxygenase reactions [Ullrich and Diehl, 1971; Poulson et al., 1974b]. Measurement of NADPH and oxygen consumption are methods of choice for work with the pure enzyme, but are complicated when using tissue fractions containing the cytochrome P450 monooxygenase. The significant endogenous utilization of NADPH or oxygen by microsomal protein fractions requires that the membrane preparations contain appreciable amounts of enzyme activity and that good substrates which have high rates of metabolism must be selected if the general methods of NADPH oxidation [Eastbrook et al., 1972] or oxygen utilization [Prough and Ziegler, 1977] are used. In addition, spectrophotometers used in measuring NADPH oxidation must be capable of accurately measuring absorbance changes in turbid samples [Estabrook et al., 1972]; oxygraphs must possess high sensitivity to small changes in oxygen concentration and have high response times [Estabrook, 1967].

The best substrate for use in measuring NADPH oxidation or oxygen utilization catalyzed by the FAD-containing monooxygenase is methimazole (2-mercapto-*N*-methylimidazole). At concentrations of 0.2–1.0 m*M* methimazole and pH 8.3, the FAD-containing monooxygenase-catalyzed S-oxidation reaction is saturated (i.e. the reaction proceeds at near V_{max}) and no appreciable further metabolism of the thionosulfur compound occurs [Poulson et al., 1974a; Poulson and Ziegler, 1979]. Glutathione can also be added to protect the enzyme from the reactive sulfenic acid product formed. The reaction can further be optimized by addition of 1 m*M* *n*-octylamine which causes a 1.4. to 1.8-fold increase in activity and simultaneously inhibits the cytochrome P450-dependent monooxygenase. However, mitochondrial contamination can lead to erroneously high rates of O_2 utilization due to the metabolism of *n*-octylamine by the mitochondrial monoamine oxidase. One must also be cautious not to preincubate the enzyme for appreciable times at 37 °C in the absence of $NADP^+$ or NADPH due to the heat lability of the enzyme under these conditions. *N,N*-Dimethylaniline may also be used as a substrate, but it is highly volatile and at pH 7–8 is not completely miscible in water at 1 m*M* concentrations (required for saturation of the enzyme).

Ziegler and co-workers have developed specific assays to measure the N-oxides and N-hydroxy compounds formed. These methods have allowed studies on the stoichiometry and the chemistry of the products of N-oxidation. *N,N*-Dimethylaniline N-oxide production can be measured as described by Ziegler and Pettit [1964]. The assay involves extraction of *N,N*-dimethylaniline, N-monomethylaniline, and aniline from the aqueous phase with ether. The N-oxide product in the water phase can be reduced back to aniline and nitrosated to give the yellow *p*-nitroso derivative. The nitroso derivative has an appreciable absorbance at 420 nm ($E_M = 8{,}200\ M^{-1}\ cm^{-1}$) and the N-oxide formed can be determined based on the amount of resultant *p*-nitroso compound measured. In addition, Mitchell and Ziegler [1969] have developed a microanalytical method to measure *N,N*-dimethylalkyl and N-methylheterocyclic amine oxides. *Tert.-N*-methyl amine oxides react with SO_2 to form a sulfite addition product which hydrolyzes in aqueous acidic solutions to form the secondary amine and formaldehyde. The formaldehyde can subsequently be measured using the method of Nash [1953]. In addition, certain N-methyl-N-hydroxylamines decompose in the presence of acids, such as 1% trichloroacetic acid, to give formaldehyde [Prough and Ziegler, 1977]. Other substrates require specific chromatographic procedures which normally must be developed for each compound of interest. A convenient, specific high performance liquid chromatography assay for *N,N*-dimethylaniline *N*-oxide has been described by Houdi and Damani [1985].

Species and Developmental Aspects

Chaykin and Block [1959] and Baker and co-workers [1962, 1963] provided the first evidence of species differences in trimethylamine N-oxide synthesis activity. They concluded that, in animal species, the activity was restricted to vertebrates. While a number of fish had the activity, little evidence was available to demonstrate trimethylamine N-oxygenase activity in plants. Mammals have been shown to contain high levels of the FAD-containing monooxygenase in liver [Baker et al., 1963; Ziegler and Pettit, 1966]. The mammalian species with high activity include guinea pig, hamster, hog, and human [Gold and Ziegler,

1973; Ziegler, 1978, 1980]. Within a species such as hog or rabbit, the liver, lung and kidney contain the highest amounts of the enzyme. It is of interest that in the rabbit and rat lung, the N-oxygenase activity is high relative to liver. Devereux and Fouts [1981] have measured the N-oxygenase activity of two purified cell types in rabbit lung. The Clara cells contained higher amounts (65%) of the N-oxygenase relative to the type II alveolar cells; cytochrome P450 is more concentrated in the Clara cell [Devereux et al., 1985].

Uehleke [1971] and Ziegler [1978] have demonstrated that the level of the FAD-containing monooxygenase in liver is reasonably low at birth and rises rapidly until weaning. Following a decrease in activity after weaning, there is a gradual increase in liver activity during adult life. In weanling rats, activity is approximately twice that seen in adults [Das and Ziegler, 1970]. The developmental pattern of the FAD-containing monooxygenase is very distinct compared to the cytochrome P450-dependent monooxygenase in rats [Uehleke, 1971]. It is of interest that there is a rapid rise in activity following birth in hamster, mouse, and rat [Ziegler, 1978]. In human fetal liver, a similar rapid rise in liver activity has been reported [Rane, 1973]. The control of these development patterns and the role of the monooxygenase during development are not known. Studies in mice have shown a sex difference in that the enzyme levels are higher in adult female mice than in males [Wirth and Thorgeirsson, 1978]. Ziegler [1980] has noted a similar situation in the hog; i.e. the enzyme activity in female liver is higher than in males. In addition, Heinze et al. [1970] studied the changes in activity of pig corpus lutea during estrous cycle. N-Oxygenase activity was low at ovulation and increased 3- to 4-fold during the first 10 to 12 days. Subsequently, the activity decreased. Under these conditions, other C-hydroxylation reactions associated with cytochrome P450 could not be measured.

The FAD-containing monooxygenase has been shown not to be inducible in hog [Masters and Ziegler, 1971], mice [Wirth and Thorgeirsson, 1978] or rats [Prough and Ziegler, 1977] by agents such as barbiturates or polycyclic aromatic hydrocarbons which are known to induce cytochrome P450 in vivo [Sato and Omura, 1978]. The sex difference noted in mice [Wirth and Thorgeirsson, 1978] and the changes in activity of corpus lutea microsomes [Heinze et al., 1970] suggest that induction of the activity may be controlled by certain steroids. Indeed, it has been reported that prednisolone induces microsomal *N,N*-dimethylaniline N-oxidase activity in rat liver [Arrhenius, 1969] and rabbit lung [Devereux and Fouts,

1974]. Duffel et al. [1981] have shown that steroids, testosterone and progesterone can modulate the FAD-containing monooxygenase in mouse liver and kidney. Castration caused the activity in liver to increase and additions of the male hormone, testosterone, to castrated males decreased the activity in liver to near control levels. Testosterone depressed liver N-oxygenase activity in gonadectomized female mice. Progesterone increased the activity in liver of castrated male mice in a manner similar to the changes noted during development of male and female mice [Wirth and Thorgeirsson, 1978]. No effect of steroid treatment was noted on mouse lung N-oxygenase activity, but testosterone increased this activity in mouse kidney tissue. Williams et al. [1985] have shown that the lung form of the enzyme appears to be under hormonal control during gestation in rabbits. It has not been established whether this new form of the enzyme exists in lung of other species or not.

Criteria for Involvement in Reactions

The initial reports by Ziegler and Pettit [1966] of the existence of a flavoprotein enzyme capable of N-oxygenating *sec.*- and *tert.*-amines which was distinct from the cytochrome P450-dependent enzyme system were not appreciated due to the difficulties both in assaying the enzyme and its heat lability. Arrhenius [1968, 1969] later demonstrated that the ratio of *N,N*-dimethyl-aniline N-oxygenase activity to N-demethylase activity increases upon treatment of liver microsomes with sodium cholate or sulfhydryl reagents like N-ethylmaleimide and methylmercury salts. Uehleke [1971] and Uehleke et al. [1971] also showed differences between the dealkylation and N-oxygenation reactions when the microsomal fractions were preincubated at 37 °C for various times in the absence and presence of either NADPH, sodium deoxycholate, or a specific inhibitor of cytochrome P450, metyrapone. In addition, Hlavica and Kiese [1969] showed differences in the apparent K_m for oxygen for the *p*-hydroxylation of acetaniline and N-hydroxylation of aniline relative to those of N-ethylaniline N-hydroxylation and *N,N*-dimethylaniline N-oxygenation. Additional differences in the pH-rate profile and induction by phenobarbital were noted. These results prompted the study of Prough and Ziegler [1977] to develop inhibitors to determine the relative participation of the two mammalian microsomal monooxygenases, FAD-containing and cytochrome-P450-containing, in the metabolism of *sec.*- and *tert.*-*N*-methylamine compounds. Prough and Ziegler [1977] have demonstrated that for *N,N*-dimethylaniline and benzphetamine the FAD-containing monooxygenase can account for better than 85% of the N-oxide synthesis while cytochrome P450 was responsible for greater than 90% of the N-demethylation reaction. In addition, Gorrod [1978a, b] has published data supportive of this conclusion.

While a number of specific inhibitors exist for cytochrome P450, few compounds exist to serve as unique inhibitors of the flavoprotein N-oxidase. For example, two groups have used immunological techniques to elicit antibodies [Patton et al., 1980; Masters and Prough, unpublished], but the antibodies produced were not inhibitory to the N-oxygenase reactions studied. However, the immunoglobulins could precipitate the purified enzyme and the microsomal membranes containing the enzyme. Only methimazole, a substrate which does not greatly interfere with the assay of *sec.*- and *tert.*-amine N-oxidase activity, and *n*-octylamine, an activator of the N-oxygenase and inhibitor of cytochrome P450 activity, can be used as biochemical tools. Table II shows some of the differences which can be utilized to distinguish and quantitate the roles of the flavin- and cytochrome-dependent monooxygenases of liver microsomes in various N-oxidation reactions. Since Hlavica and Kehl [1977] have provided evidence that a partially purified preparation of rabbit hepatic cytochrome P450 can catalyze the N-oxygenation of *N,N*-dimethylaniline at low rates, considerable effort must be expended to evaluate the relative participation of both enzymes in N-oxygenating various nitrogenous and sulfur compounds.

Table III demonstrates the effects of inhibitors on *N,N*-dimethylaniline and benzphetamine N-oxygenation and N-demethylation reactions. A specific antibody to NADPH-cytochrome *c* (P450) reductase inhibited both N-demethylation reactions by 80–90%, suggesting that these reactions are uniquely catalyzed by cytochrome P450. Conversely, the inhibitory antibody had no effect on N-oxygenation of these two *tert.*-amines. Methimazole inhibited N-oxygenation, but not N-demethylation reactions. *n*-Octylamine, the activator of the FAD-containing monooxygenase, increased the rates of N-oxide synthesis (135% of control) and decreased the rates of N-demethylation (9% of control). The judicious use of these inhibitors serves to demonstrate clearly that both enzymes simultaneously metabolize the *tert.*-amines; cytochrome P450 catalyzes N-dealkylation reactions and the FAD-containing monooxygenase catalyzes at least 85% of the N-oxygenation reactions.

Table II. Comparison of the cytochrome-P450- and FAD-containig monooxygenases.

Parameter	Cytochrome P450	FAD-containing monooxygenase
pH optima[a]	7.6	8.3
Prosthetic group	18–20 nmol heme/mg	15.2 nmol FAD/mg
Molecular weight	48,000–58,000	56,000
Multiple forms	numerous	two
Inducers in vivo	barbiturates, polycyclic hydrocarbons, TCDD, polyhalogenated aromatic hydrocarbons	testosterone prednisolone
Activators	acetone flavones	*n*-octylamine N-alkylguanidines
Inhibitors	metyrapone, *n*-octylamine, anti-NADPH cytochrome *c* reductase globulin[b], CO, SKF-525A, detergents	heat, methimazole

[a] The pH optima shown are for the N-dealkylation (cytochrome P450 monooxygenase) and N-oxidation (FAD-containing monooxygenase) reactions for *N,N*-dimethylaniline.
[b] The antibody to NADPH-cytochrome *c* (P450) reductase is an effective inhibitor since it prevents the flavoprotein reductase from donating electrons to many of its acceptors, including cytochrome P450 [Masters and Ziegler, 1971].

However, one must be cautious when using methimazole as an inhibitor since the N-oxidation of azo compounds to azoxy compounds [Fiala et al., 1977; Dunn et al., 1979] and *S*-oxygenation of some sulfur compounds [Hunter and Neal, 1975] are uniquely sensitive to inhibition by methimazole and other thiono sulfur compounds. Metabolites of methimazole [Kedderis and Rickert, 1983] and of diethyldithiocarbamate [Miller et al., 1983] appear to specifically inactivate cytochrome P450. Considerable evidence exists documenting that azoxy formation and some *S*-oxygenation reactions are, indeed, catalyzed by the cytochrome P450-dependent monooxygenase. Most likely specific forms of cytochrome P450 must exist which catalyze the N- and *S*-oxygenation reactions. Aside from the report of Hlavica and Kehl [1977] and Wiebkin and Prough [1980], little is known about this form of cytochrome P450. However, the data in table III clearly demonstrate that N-oxygenation of *tert.-N*-methyl amines catalyzed by cytochrome P450 must represent only a small proportion ($<10\%$) of the total N-oxygenation seen in the

Table III. Effect of various monooxygenase inhibitors on the N-oxidation of benzphetamine and *N,N*-dimethylaniline by liver microsomes[a]

Substrate	Inhibitor[b]	N-Demethylation[c]	N-Oxidation[c]
Pig liver microsomes (n = 3)[d]			
Benzphetamine	–	5.6	4.7
	anti-P450 reductase	0.9 (16)	4.6 (98)
	nonimmune globulin	5.4 (96)	4.8 (102)
	methimazole	4.5 (80)	1.3 (28)
	n-octylamine	0.5 (9)	6.4 (136)
Dimethylaniline	–	7.0	6.4
	methimazole	6.3 (90)	1.5 (23)
	n-octylamine	1.0 (9)	8.3 (130)
Hamster liver microsomes (n = 4)[d]			
Dimethylaniline	–	4.2	8.1
	methimazole	4.1 (98)	2.3 (28)
	n-octylamine	0.4 (10)	9.9 (122)
Rabbit liver microsome (n = 3)[d]			
Dimethylaniline	–	6.4	8.4
	methimazole	5.3 (83)	3.9 (46)
	n-octylamine	1.1 (17)	9.0 (107)
Rat liver microsomes (n = 3)[d]			
Dimethylaniline	–	8.9	5.4
	methimazole	7.6 (85)	2.4 (44)
	n-octylamine	1.2 (14)	6.3 (117)

[a] The N-demethylation and N-oxidation reactions were performed at pH 8.3 and 37 °C [Prough and Ziegler, 1977] using the methods of Nash [1953], Ziegler and Pettit [1964], and Ziegler and Mitchell [1972]. The buffer used was 0.05 *M* potassium phosphate: 0.05 *M* Tricine or 0.1 *M* Tris HC1. Similar results were seen with rat and pig microsomal fractions using either buffer system. The substrate concentrations for benzphetamine and *N,N*-dimethylaniline were 0.8 and 1.25 m*M*, respectively.
[b] The concentrations of inhibitors used were methimazole (2.5 m*M*) and *n*-octylamine (5 m*M*). The ratio of globulin protein to microsomal protein was 5 mg : 1 mg, respectively.
[c] The rates of metabolism are expressed as nmol product formed per min per mg microsomal protein. The values in parantheses are expressed as percent of control activity.
[d] The standard deviation for these means are equal to approximately 15% of the mean for pig liver microsomes and to 10% for rodent liver microsomes.

Table IV. Effect of pH on N-demethylation and N-oxidation of *N,N*-dimethylaniline by mammalian liver microsomal fractions

Species	Rate[a], nmol/min/mg			
	N-dealkylation		N-oxydation	
	pH 7.5	pH 8.3	pH 7.5	pH 8.3
Hamster	6.3	4.2	1.2	8.1
Pig	7.0	6.8	0.9	6.4
Rabbit	5.2	4.5	1.4	8.4
Rat	6.7	5.1	0.7	5.4

[a] The reactions were performed at 37 °C with 2 mg/ml liver microsomal protein and 1.25 m*M N,N*-dimethylaniline. Formaldehyde was measured by the method of Nash [1953] and *N,N*-dimethylaniline N-oxide was measured by the method of Ziegler and Pettit [1964].

presence of mammalian liver microsomes. Table IV demonstrates the differences in pH-rate optima of N-demethylation and N-oxygenation reactions catalyzed by liver microsomes. In the pig, the pH optima for *N,N*-dimethylaniline N-demethylation and N-oxygenation are approximately 7.5 and 8.3, respectively.

Conclusion

Nearly two decades have passed since Ziegler and Pettit [1966] first described the microsomal FAD-containing monooxygenase. During this time, its importance has been established based on the observations of Bickel [1969], Jenner [1971] and Ziegler [1980, 1984] that a large number of nitrogenous compounds (drugs) are excreted as N-oxides and that this enzyme system simultaneously functions with the cytochrome-P450-dependent monooxygenase in the metabolism of nitrogenous compounds [Prough and Ziegler, 1977]. In vivo, the products of the N-oxygenation reactions catalyzed by this flavoprotein monooxygenase have been underestimated due to the subsequent oxidative and reductive metabolism of the N-oxidized products. Chemically and enzymatically, many of the N-oxides and N-hydroxy compounds can be readily reduced back to the parent amine or can be oxidized to nitrones and C-nitroso compounds which are very unstable [Kadlubar et al., 1973; Kadlubar and Ziegler, 1974; Dehner et al., 1968]. In addition, some evidence exists indicating that the N-oxides can be dealkylated in vitro to give the dealkylated amine and the corresponding aldehyde [Machinist et al., 1966]. It has become clear that while the arylamides and arylamines are N-oxygenated to reactive chemical species [Pelroy and Gandolfi, 1980; Frederick et al:, 1985], many of the metabolic products of nitrogenous compounds formed in the presence of the cytochrome- and flavin-dependent monooxygenases are not toxic and these reactions may serve as detoxification reactions. Recently, Kloss et al. [1982] have suggested that the FAD-containing monooxygenase may convert norcocaine to N-hydroxynorcocaine and may catalyze a key step in cocaine-induced hepatotoxicity. Krieter et al. [1984] have demonstrated that the enzyme may enhance the biliary excretion of GSSG due to the extreme chemical reactivity of the formamidine sulfinate metabolites of thiocarbamide derivatives with reduced glutathione to form oxidized glutathione. Finally, the FAD-containing monooxygenase appears to metabolize the street drug, 1-methyl-4-phenyl-1,2,3,6-tetrahydropyridine, to its N-oxide, a potentially important metabolite [Cashman and Ziegler, 1986].

A paradoxical question exists regarding the microsomal FAD-containing monooxygenase. While it is clear that many xenobiotic compounds are biotransformed via this enzyme, no clear endogenous function has been established definitively for the N-oxygenase. Cytochrome P450, on the other hand, is known to be involved in the oxidative metabolism of steroids and long chain fatty acids [Sato and Omura, 1978]. The high concentrations of the FAD-containing monooxygenase in liver, lung and thyroid [Ziegler, 1980, 1984] suggest that some physiological role may exist for the enzyme. To date, only thiol oxidation of cysteamine and certain proteins [Poulsen and Ziegler, 1977; Ziegler and Poulsen, 1977] has been demonstrated. The significance of this enzyme in disulfide formation remains a question to be addressed in the years ahead.

Acknowledgments

The author wishes to thank D. M. Ziegler for his encouragement in writing this review and in the studies on hydrazine metabolism. Portions of the work presented were supported by grants from the American Cancer Society (BC-336) and The Robert A. Welch Foundation (I–616). The author thanks Ms. S. Dougherty for her assistance in the preparation of this chapter.

References

Arrhenius, E.: Effects on hepatic microsomal N- and C-oxygenation of aromatic amines by in vivo corticosteroid or aminofluorene treatment, diet, or stress. Cancer Res. *28:* 264–273 (1986).

Arrhenius, E.: Effects of various in vitro conditions on hepatic microsomal N- and C-oxygenation of aromatic amines. Chem.-Biol. Interact. *1:* 361–380 (1969).

Baker, J.R.; Chaykin, S.: The biosynthesis of trimethylamine-N-oxide. J. biol. Chem. 237: 1309–1313 (1962).

Baker, J.R.; Struempler, A.; Chaykin, S.: A comparative study of trimethylamine-N-oxide biosynthesis. Biochim. biophys. Acta *71:* 58–64 (1963).

Ball, S.; Bruice, T.C.: 4a-Hydroperoxyflavin N-oxidation of tertiary amines. J. Am. chem. Soc. *101:* 4017–4019 (1979).

Ball, S.; Bruice, T.C.: Oxidations of amines by a 4a-hydroperoxyflavin. J. Am. chem. Soc. *102:* 6498–6503 (1980).

Beaty, N.B.; Ballou, D.P.: Transient kinetic study of liver microsomal FAD-containing monoxygenase. J. biol. Chem. *255:* 3817–3819 (1980).

Beaty, N.B.; Ballou, D.P.: The reductive half-reaction of liver microsomal FAD-containing monooxygenase. J. biol. Chem. *256:* 4611–4618 (1981a).

Beaty, N.B.; Ballou, D.P.: The oxidative half-reaction of liver microsomal FAD-containing monooxygenase. J. biol. Chem. *256:* 4619–4625 (1981b).

Beckett, A.H.: Metabolic activation of aliphatic basic nitrogen atoms and their α-carbon atoms. Xenobiotica *1:* 365–383 (1971).

Bickel, M.H.: The pharmacology and biochemistry of N-oxides. Pharmacol. Rev. *21:* 325–355 (1969).

Cashman, J.R.; Ziegler, D.M.: Contribution of N-oxygenation to the metabolic of MPTP (1-methyl-4-phenyl-1,2,3,6-tetrahydropyridine) by various liver preparations. Molec. Pharmacol. *29:* 163–167 (1986).

Chaykin, S.; Bloch, K.: The metabolism of nicotinamide-N-oxide. Biochim. biophys. Acta *31:* 213–216 (1959).

Cleland, W.W.: The statistical analysis of enzyme kinetic data. Adv. Enzymol. *29:* 1–32 (1967).

Coomes, M.W.; Prough, R.A.: The mitochondrial metabolism of 1,2-disubstituted hydrazines, procarbazine and 1,2-dimethylhydrazine. Drug Metab. Disposit. *11:* 550–555 (1983).

Coutts, R.T.; Beckett, A.H.: Metabolic N-oxidation of primary and secondary aliphatic medicinal amines. Drug Metab. Rev. *6:* 51–104 (1977).

Cramer, J.W.; Miller, J.A.; Miller, E.C.: N-hydroxylation: a new metabolic reaction observed in the rat with the carcinogen 2-amino fluorene. J. biol. Chem. *235:* 885–888 (1960).

Das, M.L.; Ziegler, D.M.: Rat liver oxidative N-dealkylase and N-oxidase activities as a function of animal age. Archs. Biochem. Biophys. *140:* 300–306 (1970).

Dehner, E.W.; Machinist, J.M.; Ziegler, D.M.: Effect of SKF 525-A on the excretion of amine oxides by the rat. Life Sci. *7:* 1135–1146 (1968).

Devereux, T.R.; Diliberto, J.J.; Fouts, J.R.: Cytochrome P-450 monooxygenase, epoxide hydrolase and flavin monooxygenase activities in Clara cells and alveolar type II cells isolated from rabbit. Cell Biol. Toxicol. *1:* 57–65 (1985).

Devereux, T.R.; Fouts, J.R.: N-Oxidation and demethylation of *N,N*-dimethylaniline by rabbit liver and lung microsomes. Effects of age and metals. Chem.-Biol. Interact. *8:* 91–105 (1974).

Devereux, T.; Fouts, J.: Xenobiotic metabolism by alveolar type II cells isolated from rabbit lung. Biochem. Pharmacol. *30:* 1231–1237 (1981).

Doerge, D.R.; Corbett, M.D.: Hydroperoxyflavin-mediated oxidations of organosulfur compounds. Model studies for the flavin monooxygenase. Molec. Pharmacol. *26:* 348–352 (1984a).

Doerge, D.R.; Corbett, M.D.: Primary arylamide oxidation by a flavinhydroperoxide. A study of the basis for the substrate specificity of the flavoprotein monooxygenase. Biochem. Pharmacol. *22:* 3615–3619 (1984b).

Duffel, M.W.; Graham, J.M.; Ziegler, D.M.: Changes in dimethylaniline N-oxidase activity of mouse liver and kidney induced by steroid sex hormones. Molec. Pharmacol. *19:* 134–139 (1981).

Dunn, D.L.; Lubet, R.A.; Prough, R.A.: The oxidative metabolism of N-isopropyl-α-(2-methylhydrazino)-*p*-toluamide hydrochloride (procarbazine) by rat liver microsomes. Cancer Res. *39:* 4555–4563 (1979).

Entsch, B.; Ballou, D.P.; Massey, V.: Flavin oxygen derivatives involved in hydroxylation by *p*-hydroxybenzoate hydroxylase. J. biol. Chem. *251:* 2550–2563 (1976).

Erikson, J.M.; Prough, R.A.: Oxidative metabolism of some hydrazine derivatives by rat liver and lung tissues. J. Biochem. Toxicol. *1:* 1–8 (1986).

Estabrook, R.W.: Mitochondrial respiratory control and the polarographic measurement of ADP : O ratios; in Estabrook, Pullman, Methods in enzymology, vol. 10, pp. 41–47 (Academic Press, New York 1967).

Estabrook, R.W.; Cooper, D.Y.; Rosenthal, O.: The light reversible carbon monoxide inhibition of the steroid C21-hydroxylase system of the adrenal cortex. Biochem. Z. *338:* 741–755 (1963).

Estabrook, R.W.; Peterson, J.A.; Baron, J.; Hildebrandt, A.: The spectrophotometric measurement of turbid suspensions of cytochromes associated with drug metabolism. Methods in Pharmacol. *2:* 303–350 (1972).

Fiala, E.S.; Bobotas, G.; Kulakis, C.; Weisburger, J.H.: Inhibition of 1,2-dimethylhydrazine metabolism by disulfiram. Xenobiotica *7:* 5–9 (1977).

Frederick, C.B.; Mays, C.B.; Ziegler, D.M.; Guengerich, F.P.; Kadlubar, F.F.: Cytochrome P-450- and flavin-containing monooxygenase-catalyzed formation of the carcinogen N-hydroxy-2-aminofluorene and its covalent binding to nuclear DNA. Cancer Res. *42:* 2671–2677 (1982).

Ghisla, S.; Entsch, B.; Massey, V.; Husein, M.: On the structure of flavin-oxygen intermediates involved in enzymatic reactions. Eur. J. Biochem. *76:* 139–148 (1977).

Ghisla, S.; Hastings, J.W.; Favaudon, V.; Lhoste, J.-M.: Structure of the oxygen adduct intermediate in the bacterial luciferase reaction: ^{13}C nuclear magnetic resonance. Proc. natn. Acad. Sci. USA *75:* 5860–5863 (1978).

Gold, M.S.; Ziegler, D.M.: Dimethylaniline N-oxidase and aminopyrine N-demethylase activities of human liver tissue. Xenobiotica *3:* 179–189 (1973).

Gorrod, J.W.: Differentiation of the various types of biological oxidation of nitrogen in organic compounds. Chem.-Biol. Interact. *7:* 289–303 (1973).

Gorrod, J.W.: On the multiplicity of microsomal N-oxidase systems; in Singer, Ondarza, pp. 189–197 (Elsiever/North-Holland, Amsterdam 1978a).

Gorrod, J.W.: The current status of the pK_a concept in the differentiation of enzymic N-oxidation; in Gorrod, Biological oxidation of nitrogen, pp. 201–210 (Elsevier/North-Holland Biomedical Press, Amsterdam 1978b).

Heinze, E.; Hlavica, P.; Kiese, M.; Lipowsky, W.: N-Oxygenation of arylamines in microsomes prepared from corpus lutea of the cycle and other tissues of the pig. Biochem. Pharmac. *19:* 641–649 (1970).

Heubner, W.: Arbeiten aus dem pharmakologischen Institut zu Göttingen. 2. Reiche. 14. Studien über Methamoglobinidung. Arch. exp. Path. Pharmak. *72:* 241–281 (1913).

Hlavica, P.; Kehl, M.: Studies on the mechanism of hepatic microsomal N-oxide formation: the role of cytochrome P-450 and mixed-function amine oxidase in the N-oxidation of N,N-dimethylaniline. Biochem. J. *164:* 487–496 (1977).

Hlavica, P.; Kiese, M.: N-Oxidation of N-alkyl- and N,N-dialkylanilines by rabbit liver microsomes. Biochem. Pharmacol. *18:* 1501–1509 (1969).

Houdi, A.A.; Damani, L.A.: *N*-demethylation and *N*-oxidation of *N,N*-dimethylaniline by rat liver microsomes; in Gorrod, Damani, Biological oxidation of nitrogen in organic molecules, pp. 96–100 (Ellis Horwood Ltd, Chichester 1985).

Hunter, A.L.; Neal, R.A.: Inhibition of hepatic mixed-function oxidase activity in vitro and in vivo by various thiono-sulfur containing compounds. Biochem. Pharmacol. *24:* 2199–2205 (1975).

Irving, C.C.: Metabolic activation of N-hydroxy compounds by conjugation. Xenobiotica *1:* 387–398 (1971).

Jenner, P.: The role of nitrogen oxidation in the excretion of drugs and foreign compounds. Xenobiotica *1:* 399–418 (1971).

Kadlubar, F.F.; McKee, E.M.; Ziegler, D.M.: Reduced pyridine nucleotide-dependent N-hydroxy amine oxidase and reductase activities of hepatic microsomes. Archs Biochem. Biophys. *156:* 46–57 (1973).

Kadlubar, F.F.; Miller, J.A.; Miller, E.C.: Hepatic metabolism of N-hydroxy-N-methyl-4-aminobenzene and other N-hydroxyarylamines to reactive sulfuric acid esters. Cancer Res. *36:* 2350–2359 (1976).

Kadlubar, F.F.; Ziegler, D.M.: Properties of a NADH-dependent N-hydroxy amine reductase isolation from pig liver microsomes. Archs Biochem. Biophys. *162:* 83–92 (1974).

Kedderis, G.L.; Richert, D.E.: Inhibition of the microsomal *N*-hydroxylation of 2-amino-6-nitrotoluene by a metabolite of methimazole. Biochem. biophys. Res. Commun. *113:* 433–438 (1983).

Kemal, C.; Bruice, T.C.: Simple synthesis of a 4a-hydroperoxy adduct of a 1,5-dihydroflavine: preliminary studies of a model for bacterial luciferase. Proc. natn. Acad. Sci. USA *73:* 995–999 (1976).

Kiese, M.; Uehleke, H.: Der Ort der N-Oxydation des Anilins in höheren Tier. Naunyn-Schmiedebergs, Arch. exp. Path. Pharmak. *242:* 117–129 (1961).

Kloss, M.W.; Cavagnaro, J.; Rosen, G.M.; Rauckman, E.J.: Involvement of FAD-containing monooxygenase in cocaine-induced hepatotoxicity. Toxicol. appl. Pharmacol. *64:* 88–93 (1982).

Krieter, P.A.; Ziegler, D.M.; Hill, K.E.; Burk, R.F.: Increased biliary GSSG efflux from rat livers perfused with thiocarbamide substrates for the flavin-containing monooxygenase. Molec. Pharmacol *26:* 122–127 (1984).

Lintzel, W.: Untersuchungen über Trimethylammoniumbase. III. Trimethylammoniumbasen im menschlichen Harn. Biochem. Z. *273:* 243–261 (1934).

Machinist, J.M.; Dehner, E.W.; Ziegler, D.M.: Microsomal oxidases. III. Comparison of species and organ distribution of dialkylarylamine N-oxide dealkylase and dialkylarylamine N-oxidase. Archs Biochem. Biophys. *125:* 858–864 (1968).

Machinist, J.M.; Orme-Johnson, W.H.; Ziegler, D.M.: Microsomal oxidases. II. Properties of a pork liver microsomal N-oxide dealkylase. Biochemistry *5:* 2939–2943 (1966).

Masters, B.S.S.; Ziegler, D.M.: The distinct nature and function of NADPH-citochrome *c* reductase and the NADPH-dependent mixed-function amine oxidase of porcine liver microsomes. Archs Biochem. Biophys. *145:* 358–364 (1971).

Miller, G.E.; Zemaitis, M.A.; Greene, F.E.: Mechanisms of diethyldithiocarbamate-induced loss of cytochrome P-450 from rat liver. Biochem. Pharmacol. *32:* 2433–2439 (1983).

Mitchell, C.H.; Ziegler, D.M.: A quantitative micro method for the estimation of amine oxides. Analyt. Biochem. *28:* 261–268 (1969).

Nash, T.: The colorimetric estimation of formaldehyde by means of the Hantzsch reaction. Biochem. J. *55:* 416–421 (1953).

Ohmiya, A.; Mehendale, H.M.: Comparison of pulmonary *N*-oxidation of chlorpromazine, dimethylaniline, and imipramine in rabbit and rat. Res. Commun. Psychol. Psychiat. Behav. *6:* 205–212 (1981).

Omura, T.; Sato, R.: The carbon monoxide-binding pigment of liver microsomes. I. Evidence for its hemoprotein nature. J. biol. Chem. *239:* 2270–2285 (1964).

Patton, S.E., Rosen, G.M.; Rauckman, E.J.; Graham, D.G.; Small, B.; Ziegler, D.M.: Hamster hepatic nuclear mixed-function amine oxidase: location and specific activity. Molec. Pharmacol. *18:* 151–156 (1980).

Pelroy, R.A.; Gandolfi, A.J.: Use of a mixed-function amine oxidase for metabolic activation in the Ames/salmonella assay system. Mutat. Res. *72:* 329–334 (1980).

Poulsen, L.L.; Hyslop, R.M.; Ziegler, D.M.: S-Oxidation of thioureylenes catalyzed by a microsomal flavoprotein mixed-function oxidase. Biochem. Pharmac. *23:* 3431–3440 (1974a).

Poulsen, L.L.; Hyslop, R.M.; Ziegler, D.M.: S-Oxygenation of N-substituted Thioureas catalyzed by the pig liver microsomal FAD-containing monooxygenase. Archs Biochem. Biophys. *198:* 78–88 (1979).

Poulsen, L.L.; Kadlubar, F.F.; Ziegler, D.M.: Role of the microsomal mixed-function amine oxidase in the oxidation of *N,N*-disubstituted hydroxylamines. Archs Biochem. Biophys. *164:* 774–775 (1974b).

Poulsen, L.L.; Masters, B.S.S.; Ziegler, D.M.: Mechanism of 2-naphthylamine oxidation catalyzed by pig liver microsomes. Xenobiotica *6:* 481–498 (1976).

Poulsen, L.L.; Ziegler, D.M.: Microsomal mixed-function oxidase-dependent renaturation of reduced ribonuclease. Archs Biochem. Biophys. *183:* 563–570 (1977).

Poulsen, L.L.; Ziegler, D.M.: The liver microsomal FAD-containing monoxygenase. Spectral characterization and kinetic studies. J. biol. Chem. *254:* 6449–6455 (1979).

Prough, R.A.: The N-oxidation of alkylhydrazines catalyzed by the microsomal mixed-function amine oxidase. Archs Biochem. Biophys. *158:* 442–444 (1973).

Prough, R.A.: Freeman, P.C.; Hines, R.N.: The oxidation of hydrazine derivatives catalyzed by the purified liver microsomal FAD-containing monooxygenase. J. biol. Chem. *256:* 4178–4184 (1981).

Prough, R.A.; Ziegler, D.M.: The relative participation of liver microsomal amine oxidase and cytochrome P-450 in N-demethylation reactions. Archs Biochem. Biophys. *180:* 363–373 (1977).

Rane, A.: N-Oxidation of a tertiary amine (*N,N*-dimethylaniline) by human fetal liver microsomes. Clin. Pharmacol. Ther.-*15:* 32–38 (1973).

Rauckman, E.J.; Rosen, G.M.; Kitchell, B.B.: Superoxide radical as an intermediate in the oxidation of hydroxylamines by mixed function amine oxidase. Molec. Pharmacol. *15:* 131–137 (1979).

Sato, R.; Omura, T.: Cytochrome P-450, pp. 1–233 (Academic Press, New York 1978).

Shannon, P.; Bruice, T.C.: A novel system for the N-dealkylation reaction. J. am. chem. Soc. *103:* 4580–4582 (1981).

Smyser, P.S.; Sabourin, P.J.; Hodgson, E.: Oxidation of pesticides by purified microsomal FAD-containing monooxygenase from mouse and pig liver. Pestic. Biochem. Physiol. *24:* 368–374 (1985).

Tabor C.W.; Tabor, H.; Rosenthal, S.M.: Purification of amine oxidase from beef plasma. J. biol. Chem. *208:* 645–661 (1954).

Tynes, R.E.; Sabourin, P.J.; Hodgson, E.: Identification of distinct hepatic and pulmonary forms of microsomal flavin-containing monooxygenase in the mouse and rabbit. Biochem. biophys. Res. Commun. *126:* 1069–1075 (1985).

Uehleke, H.: N-Oxydation von N-alkylierten Anilinen durch isolierte Organ, Zellen, und Zellbestandteile. Naunyn-Schmiedebergs, Arch. exp. Path. Pharmak. *241:* 150–151 (1960).

Uehleke, H.: N-Hydroxylation. Xenobiotica *1:* 327–338 (1971).

Uehleke, H.; Breyer, U.; Budczies, B.; Taburelli, S.; Hillmer, K.-H.: Der Einfluss von Metyrapon und verschiedene Typen von mikrosomalen N-Oxidationen. Hoppe-Seylers Z. physiol. Chem. *352:* 403–411 (1971).

Ullrich, V.; Diehl, H.: Uncoupling of monooxygenation and electron transport by fluorocarbons in liver microsomes. Eur. J. Biochem. *20:* 509–512 (1971).

Weisburger, J.H.; Weisburger, E.K.: Biochemical formation and pharmacological, toxicological, and pathological properties of hydroxylamines and hydroxamic acids. Pharmacol. Rev. *25:* 1–66 (1973).

White, R.E.; Coon, M.J.: Oxygen activation by cytochrome P-450. A. Rev. Biochem. *49:* 315–356 (1980).

Wiebkin, P.; Prough, R.A.: Oxidative metabolism of N-Isopropyl-α-(2-methylazo)-*p*-toluamide (azo-procarbazine) by rodent liver microsomes. Cancer Res. *40:* 3524–3529 (1980).

Williams, D.E.; Hale, S.E.; Muerhoff, A.S.; Masters, B.S.S.: Rabbit lung flavin-containing monooxygenase. Purification, characterization, and induction during pregnancy. Molec. Pharmacol. *28:* 381–390 (1985).

Williams, D.E.; Ziegler, D.M.; Nordin, D.J.; Hale, S.E.; Masters, B.S.S.: Rabbit lung flavin-containing monooxgenase is immunochemically and catalytically distinct from the liver enzyme. Biochem. biophys. Res. Commun. *125:* 116–122 (1984).

Wirth, P.J.; Thorgeirsson, S.S.: Amine oxidase in mice-sex differences and developmental aspects. Biochem. Pharmacol. *27:* 601–603 (1978).

Ziegler, D.M.: Microsomal oxidases; in Fleischer, Hatefi, MacLennan, Tzagloff, The molecular biology of membranes, chapt. 9, pp. 193–204 (Plenum Publishing, New York 1978).

Ziegler, D.M.: Microsomal flavin-containing monooxygenase: oxygenation of nucleophilic nitrogen and sulfur compounds; in Jakoby, Enzymatic basis of detoxication, pp. 201–227, vol, 1, chapter 9 (Academic Press, New York 1980).
Ziegler, D.M.: Metabolic oxygenation of organic nitrogen and sulfur compounds; in Mitchell, Horning, Drug metabolism and drug toxicity, pp. 33–53 (Waverly Press, Baltimore 1984).
Ziegler, D.M.; Jollow, D.; Cook, D.E.: Properties of a purified liver microsomal mixed function amine oxidase; in Kamin, Flavins and flavoproteins. 3rd Int. Symp., pp. 507–522 (University Park Press, Baltimore 1971).
Ziegler, D.M.; McKee, E.M.; Poulsen, L.L.: Microsomal flavoprotein-catalyzed N-oxidation of arylamines. Drug. Metab. Disposit. *1:* 314–321 (1973).
Ziegler, D.M.; Mitchell, C.H.: Microsomal oxidase IV. Properties of a mixed-function amine oxidase isolated from pig liver microsomes. Archs Biochem. Biophys. *150:* 116–125 (1972).
Ziegler, D.M.; Pettit, F.H.: Formation of an intermediate N-oxide in the oxidative demethylation of *N,N*-dimethylaniline catalyzed by liver microsomes. Biochem. biophys. Res. Commun. *15:* 188–193 (1964).
Ziegler, D.M.; Pettit, F.H.: Microsomal oxidases. I. The isolation and characterization of the dialkylarylamine oxygenase activity of pork liver microsomes. Biochemistry *5:* 2932–2938 (1966).
Ziegler, D.M.; Poulsen, L.L.: Protein disulfide bond synthesis: a possible intracellular mechanism. Trends. Biochem. Sci. *2:* 79–81 (1977).
Ziegler, D.M.; Poulsen, L.L.: Hepatic microsomal mixed-function amine oxidase; in Fleischer, Packer, Methods in enzymology, vol. 52C, pp. 142–151 (Academic Press, New York 1978).

R.A. Prough, PhD, Department of Biochemistry, The University of Louisville School of Medicine, Louisville, KY 40292 (USA)

Prog. bas. clin. Pharmac., vol. 1, pp. 132–160 (Karger, Basel 1988)

V. Nitrogen Oxidation in Carcinogenesis

Check Y. Quon

Clinical Pharmacology and Drug Metabolism, Research and Development, Du Pont Critical Care, Waukegan, Ill., USA

Introduction

Epidemiologists have estimated that environmental factors may be involved in the etiology of 80–90% of all human cancer [40], a large proportion of which is caused by natural or synthetic chemicals. With the exception of direct alkylating agents and a few inorganic metals, most chemical carcinogens require metabolic transformation to active metabolites which initiates the chain of complex events leading to tumor formation.

This chapter will focus on the role of *N*-oxidation in carcinogenesis, specifically the *N*-oxidation of aromatic amines, aromatic amides, and aminoazo dyes. For the sake of simplicity, *N*-nitroso compounds will not be covered. Excellent reviews are available discussing various aspects of these compounds [72, 97, 122].

Historical Background

Chemical carcinogenesis was first demonstrated in the dye industry where epidemiological studies clearly indicated that certain cancers were peculiar to small occupational groups. A correlation between aromatic amine exposure and human cancer was first reported by Rehn [117] who noted bladder cancer in workers who came in contact with aniline. Rehn called this neoplasm 'aniline' cancer because aniline was the most abundant chemical in these factories. Subsequent studies, however, identified

2-naphthylamine (2-NA) and 4-aminobiphenyl (4-AP) as causative agents [18, 19, 129].

The first successful animal model for human bladder cancer was reported by Hueper et al. [42] when these researchers induced bladder cancer by the administration of 2-NA to dogs. Soon after, liver tumors in rats were induced by ortho-aminoazotoluene or 2′,3-dimethyl-4-aminoazobenzene [154] and 4-dimethylaminoazobenzene (DAB) [59]. 2-Acetylaminofluorene (AAF), probably the most extensively studied of all chemical carcinogens, was initially patented as a highly effective insecticide until it was discovered that upon chronic feeding to rats, tumors developed in the liver and a variety of other tissues [144].

Activation Steps

N-Hydroxylation

Unlike polyaromatic hydrocarbons and alkylating agents which can induce tumors at the site of administration, aromatic amine and amide and aminoazo dye type carcinogens generally induce tumors at sites distant from their sites of entry into the body. This fact, coupled with observations that different animal species have differing susceptibilities to the carcinogenic effects of these compounds, led to the hypothesis of metabolic activation [88]. Early attempts at correlating *C*-oxidation of these compounds with carcinogenicity was not successful. Subsequently, attention was switched to *N*-oxidation as perhaps having carcinogenic significance.

Chemical oxidation of aromatic amines to their corresponding hydroxylamines is a well-established reaction. As early as 1915, it was suggested that *N*-oxidation of aromatic amines might occur in vivo [41]. Biological hydroxylation of nitrogen was first demonstrated by Cramer et al. [21] when a new urinary metabolite of AAF was isolated after feeding AAF to rats and identified as the *N*-hydroxy derivative of AAF. Table I lists the percentage of an administered dose of AAF recovered as the *N*-hydroxy metabolite in the urine of different species. In vitro studies [79] using rat liver microsomes indicated that at AAF concentrations of less than 1μM, *N*-hydroxylation is a major pathway accounting for as much of 30% of the total metabolites formed. At higher AAF concentrations, ring hydroxylation predominates and *N*-hydroxylation becomes a minor pathway of metabolism. Subsequently, many other

Table I. *N*-Hydroxylation of 2-acetylaminofluorene in several species

Species	Dose excreted in urine as *N*-hydroxy derivative, %	Reference(s)
Cat	1.5	134
Dog	5.2	109
Guinea pig	0	89
Hamster	5–20	89, 135
Man	4–14	133
Monkey	0.6–2.7	24, 26
Mouse	1.8–2.3	89, 93
Rabbit	13–20	44
Rat	0.3–15[1]	93

[1] Chronic dosage. Modified from ref. [136].

aromatic amines, aromatic amides, and aminoazo dyes were shown to undergo this metabolic pathway [15, 16, 88, 92].

It is noteworthy that *N*-hydroxy derivatives of aromatic amides like AAF are chemically more stable and therefore more amenable to isolation, whereas *N*-hydroxy derivatives of aromatic amines are less stable and prone to further oxidation, forming nitroso derivatives. Also, *N*-hydroxy derivatives of aromatic amides, called hydroxamic acids, are acidic, whereas *N*-hydroxy derivatives of aromatic amines, called hydroxylamines are basic.

Despite the varied chemical structures of carcinogenic compounds containing nitrogen, the common denominator in the carcinogenic effects of all these compounds appears to be the requirement for *N*-hydroxylation prior to tumor induction. There were several reasons to support the concept that *N*-hydroxylation is a necessary step in the carcinogenic effect of these compounds: (1) Parent compounds in general are not active at their administration sites but at locations distant from their sites of entry [14]. (2) *N*-hydroxy metabolites have been shown to be active at local sites of administration [94]. (3) *N*-hydroxy compounds are active at sites where the parent compounds are not [114]. (4) *N*-hydroxy metabolites on a molar basis are more potent than parent compounds [89, 90]. (5) Animal species that have low *N*-hydroxylating activity are not very susceptible to the carcinogenic effects of these compounds [89]. (6) Structurally similar compounds that cannot be *N*-hydroxylated are not carcinogenic [35]. (7) Parent compounds have larger fluctuations in tumor response in different species whereas tumor responses to *N*-hydroxy metabolites are more uniform [89].

Esterification

If *N*-hydroxy metabolites are the chemical species responsible for tumor formation, then these compounds must be able to interact with vital tissue components like protein, RNA, and DNA. *N*-Hydroxy derivatives of carcinogenic amines and amides were capable of alkylating tissue macromolecules at high and low pH [62] but at physiological pH, the *N*-hydroxy derivatives were inactive [3, 32, 83]. These findings indicated that *N*-hydroxylation is necessary, but not sufficient, in transforming inert compounds into reactive species. Further activation of *N*-hydroxy derivatives must occur before these metabolites can react with tissue macromolecules at physiological pH.

The current, most generally accepted hypothesis for formation of the ultimate form of chemical carcinogens of this type is that an esterification step is involved in transforming *N*-hydroxy derivatives to highly reactive species [82, 92]. This hypothesis was based on several different observations: (1) The corresponding esters are more carcinogenic than the parent *N*-hydroxy compounds from which they are derived. (2) The esters react with tissue macromolecules at physiological pH. (3) Enzymes for formation of some of the esters can be found in vivo. (4) The in vitro adducts formed by reacting ester derivatives with tissue macromolecules are similar to adducts isolated from animals administered either the parent amines or their corresponding *N*-hydroxy metabolites.

The first ester synthesized and examined was *N*-benzoyloxy-4-monomethylaminoazobenzene (*N*-benzoyloxy MAB), the benzoyl ester of N-hydroxy MAB. *N*-Benzoyloxy MAB when injected into animals was a very powerful carcinogen [110] and reacted nonenzymatically at physiological pH with various tissue nucleophiles including DNA, RNA, guanine, tyrosine, methionine, and cysteine [92, 121]. Subsequently, many other esters of aromatic amines, aromatic amides, and aminoazo dyes were synthesized and examined. For example, with AAF, the *N*-acetoxy, *N*-sulfate, *N*-benzoyloxy, *N*-propionyloxy and *N*-butyryloxy derivatives were examined, and all were reactive in vitro under physiological conditions, and all were in vivo carcinogens.

Despite the in vitro reactivity of ester forms of *N*-hydroxy derivatives, most of the synthetic esters are not anticipated to be formed in vivo with the possible exception of the sulfate, phosphate, and glucuronide conjugates. Demonstration by two different groups [22, 57] of the presence of sulfotransferases in rat liver cytosol added credence to the concept that the ultimate form of these nitrogen-containing carcinogens is the sulfuric acid ester of the *N*-hydroxy metabolite.

Liver cytosol sulfotransferases transfer activated sulfate in the form of 3′-phosphoadenosine-5′-phosphosulfate to *N*-acetyl-*N*-arylhydroxylamines. Formation of this type of ester was first demonstrated for AAF [22]. Subsequently, many other carcinogenic *N*-hydroxy compounds, including N-OH-MAB, *N*-hydroxy-4-aminoazobenzene (N-OH-AB) *N*-hydroxy-1-naphthylamine (N-OH-1-NA), and N-OH-2-NA, were also shown to be substrates for these enzymes [52].

The fact that acetanilide can inhibit the carcinogenicity of AAF [22] which can be reversed by adding sulfate ions to the diet of rats [141] coupled with the finding that pentachlorophenol, a sulfotransferase inhibitor, blocks the tumorigenic effects of AAF [81, 103] have led to the conclusion that at least in rat liver, the ultimate reactive form of AAF and possibly other structurally similar compounds is the sulfuric acid ester of N-OH-AAF (fig. 1).

The ultimate, reactive form of AAF and other carcinogens in extrahepatic tissues, however, is still in doubt. Two rat tissue sites that are very susceptible to the carcinogenic effects of AAF are the rat mammary gland and ear duct sebaceous gland (Zymbal's gland). These two tissues contain cytochrome P450 and its associated oxidative enzymes [108], but attempts to demonstrate sulfotransferase activity were unsuccessful [48]. These findings have resulted in a search for another ultimate carcinogen(s).

Ultimate carcinogens in extrahepatic tissues could be the electrophilic-*N*-acyloxyarylamines formed by the action of aromatic hydroxamic acid *N*-*O*-acyltransferases present in the cytosol of rat liver, kidney, small intestinal mucosa [4, 5, 7, 56] and mammary gland [4, 58]. The *N*-*O*-acyltransferases transfer acyl groups from the hydroxylamine nitrogen in aromatic hydroxamic acids to the oxygen of the hydroxy group, forming highly reactive *N*-acetoxy derivatives [4, 5, 7]. Reaction of *N*-acetoxy derivatives with macromolecules could explain the significant binding of aminoaryl moieties to proteins [5] and the C-8 position of guanine in nucleic acids [50, 57] after administration of aromatic amides.

Activation of hydroxylamine metabolites of carcinogenic amines can also be mediated by cytosolic 5-acetylcoenzyme-A-dependent en-

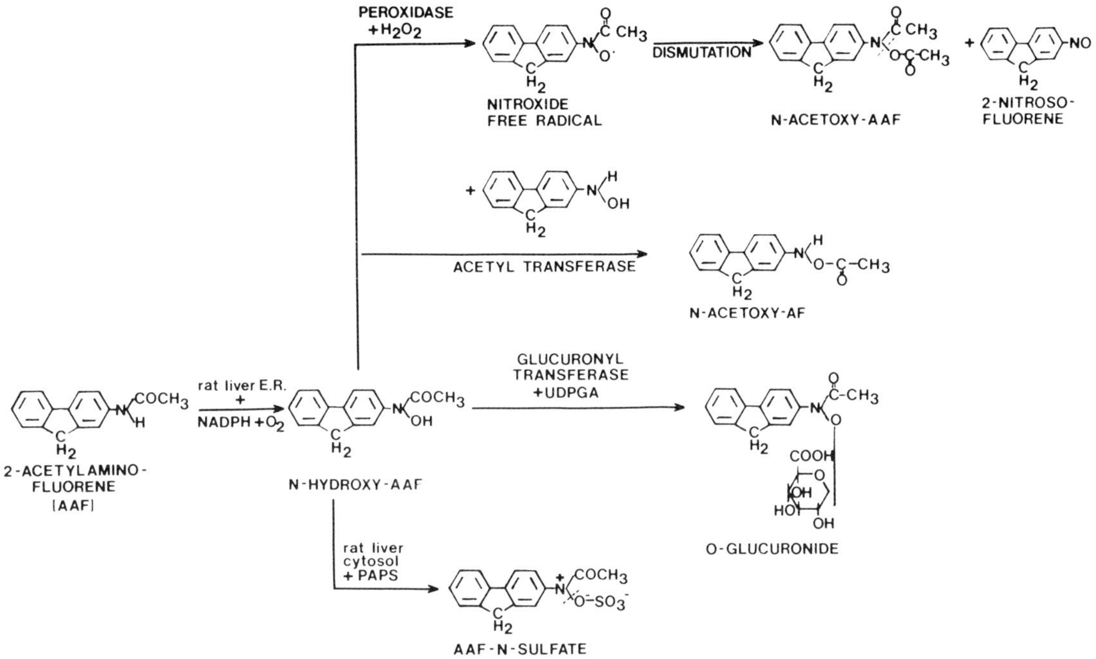

Fig. 1. Possible metabolic pathways for the activation of 2-acetylaminofluorene into reactive species. Modified from ref. [82].

zymes under neutral conditions. With rat hepatic cytosol, binding to DNA was observed with N-OH-3-2′-dimethyl-4-AP, N-OH-2-AF, N-OH-4-AP, N-OH-N′-acetylbenzidine and N-OH-2-NA but not with N-OH-N-methyl-4-AB [27]. S-Acetylcoenzyme-A-dependent activation of N-OH-3-2′-dimethyl-4-AP was detected using hepatic cytosol from rabbit, hamster, rat, human, guinea pig and mouse, but not with dog liver or bladder cytosol. Binding of N-OH-3-2′-dimethyl-4-AP to nucleic acids was not observed when arylhydroxamic acid *N-O*-acyltransferase and 3′-phosphoadenosine-5′-phosphosulfate-dependent assays were conducted with rat liver cytosol [28]. These studies indicate that the *S*-acetylcoenzyme-A-dependent formation of reactive *N*-acetoxy arylamines by cytosolic acetyltransferase(s) could serve as a major metabolic activation pathway in several species, particularly those which cannot utilize arylhydroxamic acids as acyl donors for intramolecular *N-O*-acyltransfer or for intermolecular transacetylation of N-OH-arylamines.

Other reactive forms of carcinogenic amines and amides have also been proposed for different tissue sites, among them the *N*- and *O*-glucuronide conjugates of the *N*-hydroxy metabolites. As much as 30% of a dose of AAF is excreted as the *O*-glucuronide conjugate of N-OH-AAF by rabbits [44]. The *O*-glucuronide of N-OH-AAF can react with nucleic acids or proteins [45], and the compound has been proposed to contribute to the carcinogenic effects of AAF and N-OH-AAF on the epithelial lining of the urinary tract of rabbits [46].

Similarly, *N*-glucuronide conjugates of *N*-arylhydroxylamines can be formed in vitro [53] and in vivo [54, 104, 116]. In fact, the N-glucuronide conjugates of N-OH-2-NA and N-OH-4-AP have been postulated to serve as stable transport forms for these carcinogenic aromatic amines [53]. Hydrolysis of these *N*-glucuronides to the free hydroxylamine in the normally acidic urine of dogs and man and protonation of the hydroxylamine could result in the formation of highly electrophilic nitrenium ions at or near the target site of these carcinogenic compounds (fig. 2).

A pharmacokinetic model has been developed to support the role of *N*-arylhydroxylamines in bladder tumorigenesis [155]. The model predicts that the exposure level of the bladder epithelium to N-arylhydroxylamines is dependent on urinary pH and voiding interval. Based on these two parameters, the model predicts the exposure level to be man $\geq$ dog $>$ monkey $>$ rat which is in excellent agreement with the reported carcinogenic potency of 2-NA in these species [113, 151]. In addition, this model predicts that alterations of urinary pH should result in a corresponding change in the amount of free *N*-arylhydroxylamines in urine. This was substantiated by a study in which rats were given ammonium chloride or ammonium carbonate to lower or raise, respectively, the urine pH [54]. An inverse relationship between urine pH and the ratio of free hydroxylamine to *N*-glucuronide conjugate, occurred i.e. the lower the pH, the higher the ratio and the higher the pH, the lower the ratio.

Free Radicals

It has been known for more than two decades that antioxidants can suppress the size or number of tumors produced by carcinogens of various chemical structures including aromatic amines, amides, and aminoazo dyes [128]. Antioxidants, like butylated hydroxytoluene and ascorbate, quench-free radical processes and this property of antioxidants has often been suggested to be the mechanism of tumor protection.

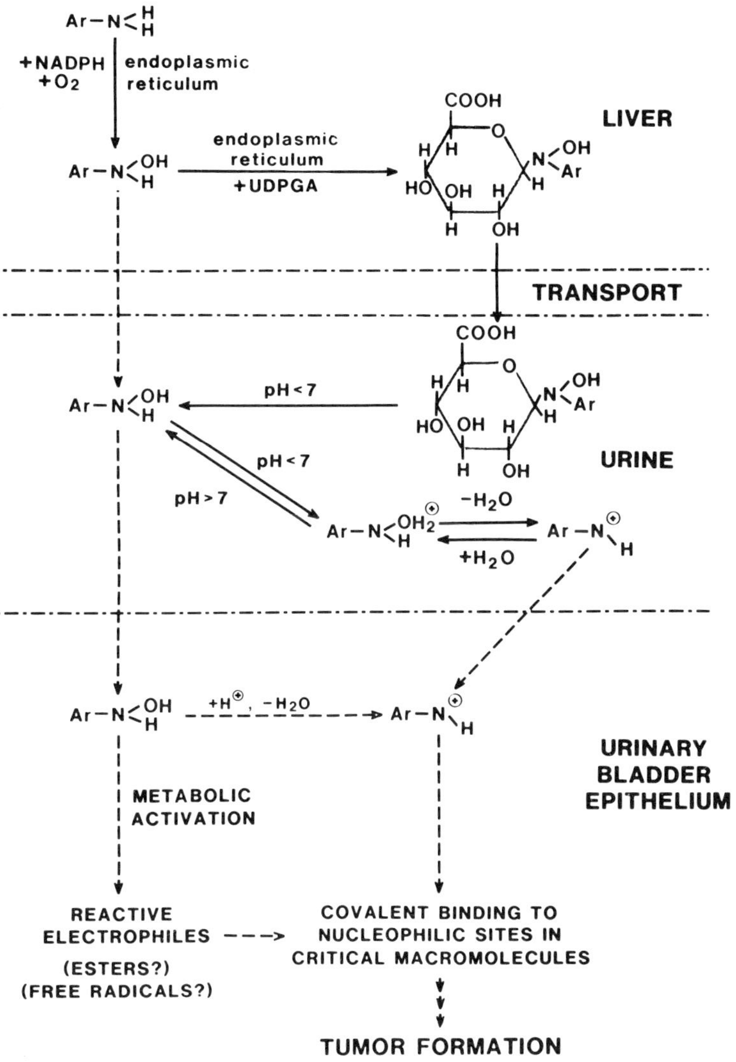

Fig. 2. Formation and transport of possible proximate and ultimate carcinogenic metabolites of arylamines for the induction of urinary bladder cancer. AR = Aryl substituent. Adapted from ref. [53].

Free radical involvement in the etiology of chemical carcinogens has been proposed as an additional mechanism for activation of *N*-hydroxy derivatives of aromatic amines and amides. In the presence of ferricyanide or silver oxide in benzene, 2 molecules of N-OH-AAF can dismutate to 2-nitrosofluorene and *N*-acetoxy AAF via a nitroxyl free radical intermediate [8, 31]. The free radical dismutation model was subsequently verified in different biological systems using horseradish peroxidase, lactoperoxidase, or human myeloperoxidases [6, 7] and various peroxides, including hydrogen peroxide, cumene hydroperoxide, and linoleic acid hydroperoxide [6, 29].

Incubation of 2-nitrosofluorene with liposomes or rat liver microsomes results in covalent binding of NOF to unsaturated membrane lipids, forming the nitroxyl free radical adduct (N-O-LAF). Indirect evidence has supported the role of N-O-LAF in aromatic amine and amide carcinogenesis [30]. First, rat liver microsomes contain enzymes which catalyze the formation of N-O-LAF from N-OH-AAF and this reaction is inhibited by antioxidants. Second, *N*-acetoxy AAF, and 2-nitrosofluorene, when incubated with rat liver microsomes, result in formation of N-O-LAF. And third, N-O-LAF from isolated microsomal membranes is mutagenic in the Ames test.

Prostaglandin H Synthase

An additional mechanism for the activation of carcinogenic amines was first demonstrated by Marnett et al. [73] when they reported the cooxidation of several chemicals, including benzo(a)pyrene, during the conversion of arachidonic acid to prostaglandins via a peroxidase like reaction. The enzyme involved, prostaglandin H synthase (PHS), is abundant in extrahepatic tissues which are low in monooxygenase activity [25].

Aromatic amines are excellent co-substrates for PHS. Ram seminal vesicle microsomes, when fortified with arachidonic acid, catalyze the dealkylation of various *N*-methyl compounds [124]. Carcinogenic aromatic amines, in particular benzidine and 2-AF, have been extensively studied as co-substrates of PHS.

Benzidine was not metabolized by either NADPH- or arachidonic-acid-dependent systems when incubated with liver microsomes [149, 157]. However, arachidonic-acid-dependent metabolism of benzidine, as measured by the formation of covalently bound metabolites, was observed in kidney medulla but not cortex microsomes [149, 156], which is consistent with PHS activity being present in the medulla but not the cortex [25, 148, 150].

Binding of benzidine to dog bladder transitional epithelial microsomal protein and DNA was dependent on the addition of arachidonic acid, blocked by indomethacin, and not dependent on NADPH. The rate of benzidine binding to macromolecules was considerably greater with bladder microsomes than liver microsomes. These results are consistent with the involvement of bladder transitional epithelial PHS in the genesis of primary amine-induced bladder cancer [149].

2-AF in the presence of arachidonic acid and a source of PHS is oxidized to nitrofluorene, 2,2′-bisazofluorene and material covalently bound to microsomal proteins [13, 102]. Formation of nitrofluorene and 2,2′-bisazofluorene from 2-AF is consistent with a one-electron mechanism of 2-AF oxidation by PHS in which the *N*-hydroxy intermediate is not involved in 2-AF metabolism by PHS [12]. PHS activated 2-AF also covalently binds to nucleic acids, but not as efficiently as benzidine [51, 102].

Structure Activity Relationships

A key requirement for any amino compound to have carcinogenic properties appears to be conformity to the 'para principle' first enunciated by Druckrey [23]: 'In order for any aromatic amine, amide, or aminoazo dye to be carcinogenic, the compound must have a long uninterrupted conjugated system with the amino group attached to one of the terminal (para) carbons.' The most active carcinogens in this group are those having two or three aromatic rings. Monocyclic aromatic amines and amides are inactive or very weak carcinogens. Aromatic amines with more than three rings are also weak carcinogens. The most active two or three ring compounds all conform to the 'para principle'.

Aromatic Amines

The two most extensively studied aromatic amines are the aminobiphenyls and naphthylamines. The position of the amino group in biphenyl appears to be a key determinant of the carcinogenic activity of this group of compounds (fig. 3). 4-AP and *4*-biphenylacetamide are moderate carcinogens in rats whereas 2, 3- and 2-biphenylacetamide are only weakly carcinogenic [91]. Substitution of a methyl group ortho to

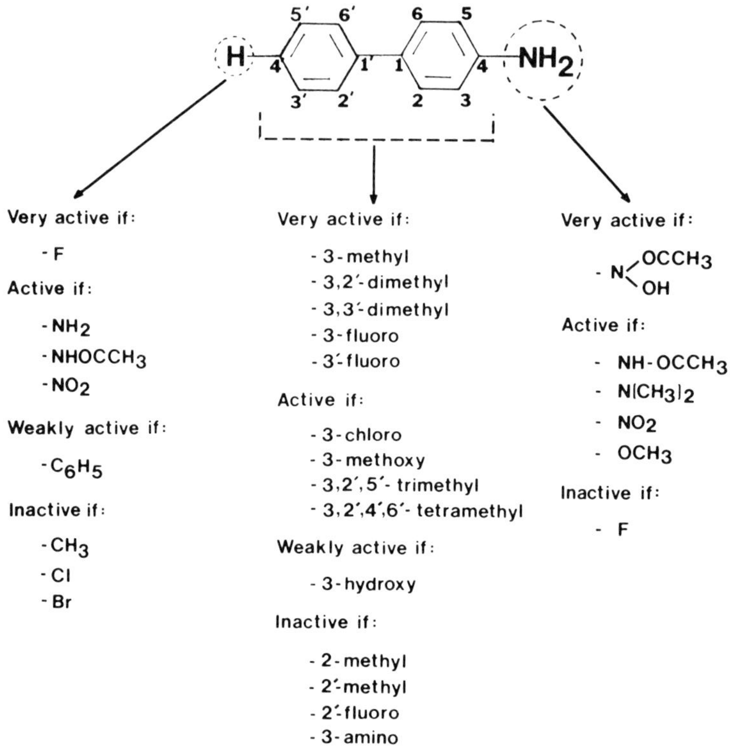

Fig. 3. Structural requirements for carcinogenecity of 4-aminobiphenyl derivatives. Modified from ref. [2].

the amino function often enhances the carcinogenicity of the unsubstituted aminobiphenyl compound, as illustrated by the potency of 3,2′-dimethylaminobiphenyl [131]. Studies showed that 3-methyl, 3,3′-dimethyl, 3,2′,5′-trimethyl, and 3,2′,4′,6′-tetramethyl derivatives are carcinogenic [130], whereas 2- and 2′-methyl aminobiphenyl and their acetylated derivatives are inactive [91, 130, 131]. Ortho substitution with a hydroxy or chlorine does not increase potency [91, 130] but substitution at the para position appears to increase potency due to blockade of detoxification mechanisms, for example, C-hydroxylation at that position [75].

The positional isomers 1-NA and 2-NA demonstrate again the importance of the location of the amine group. Early studies [11, 43] demonstrated carcinogenic activity for 1-NA, but the 1-NA samples were contaminated with 4–11% of 2-NA. Later studies [113, 115] using highly purified 1-NA indicated that the pure compound failed to induce bladder cancer after administration of moderately high doses to dogs over a 3-year period. Recent findings that hepatic microsomes from rats, dogs, and humans were capable of catalyzing the *N*-oxidation of 2-NA but not 1-NA while 1-NA can be hydroxylated at the 2 position, suggest that 1-NA is shunted towards detoxification instead of activation pathways [39].

Aromatic Amides

Importance of the amino group in the action of AF and AAF in rats was demonstrated by the lack of activity of 2-chlorofluorene, fluorene, and fluorenone (fig. 4) [145–147]. The 2- and 7- positions of fluorene are para with respect to the biphenyl linkage and therefore the observed potency of compounds substituted at these two positions is in good agreement with the 'para principle'. For any substantial carcinogenic activity there must be an amino group or its oxidized product (i.e. hydroxylamino, nitro or nitroso) in one or both of the 2- and 7- positions. Displacement of the acetylamino group from the 2- or 7- position to somewhere else on the ring results in compounds with weak or no carcinogenic activity. Thus, 1-, 3-, and 4-acetylaminofluorene are only weakly or marginally active [98, 100, 139] and 9-acetylaminofluorene is totally inactive [119]. Addition of a second acetylamino group forming 2,7-bis-acetylaminofluorene greatly enhances the carcinogenecity [101].

Since AAF is metabolically deacetylated to AF in vivo [34, 138] and in vitro [132] by a reversible reaction [105], the general impression is that AF, AAF, and 2-diacetylaminofluorene are of comparable potency and differ only in the latency period [87, 99]. Any substitution on the amino group which decreases or blocks *N*-hydroxylation decreases the potency of the compound. Thus, methylation of the amino group reduces carcinogenicity and target spectrum [87, 119].

The relative carcinogenicity of N-OH-AAF, 2-nitrosofluorene, and 2-nitrofluorene roughly reflects their closeness to the proximate carcinogen. N-OH-AAF is more potent than AAF and 2-nitrosofluorene is as potent as AAF in some target sites [38]. 2-Nitrofluorene is the least potent of all of the *N*-oxidized derivatives both in terms of tumor incidence and tissues affected [87, 98]. Halogenation in the 7 position has mixed effects depending on the substituents. Fluorination results in a

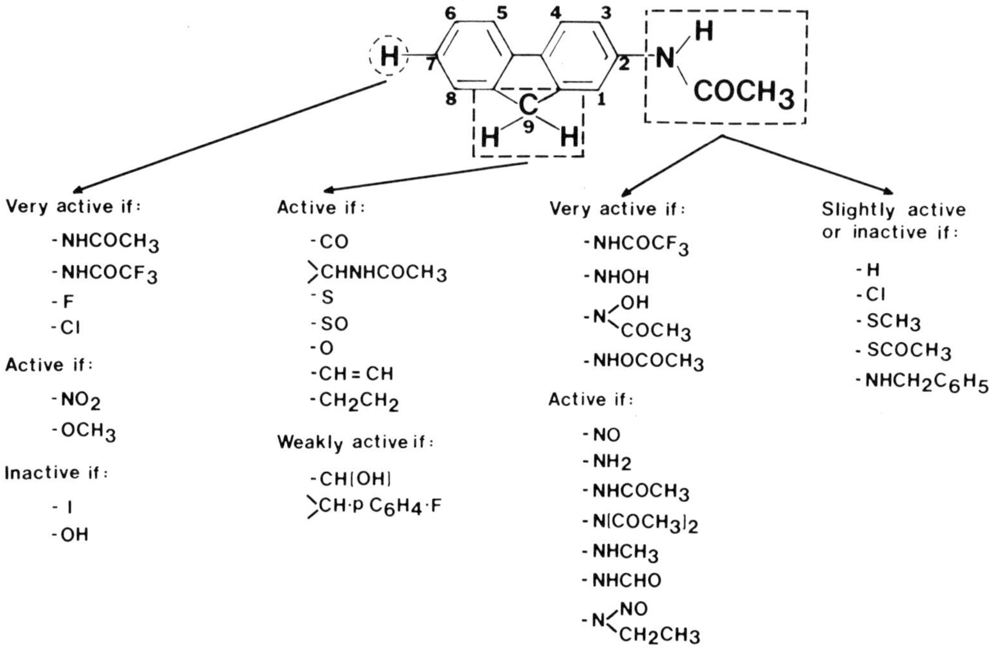

Fig. 4. Structural requirements for carcinogenecity of 2-acetyl-aminofluorenes derivatives. Modified from ref. [2].

compound that is comparable to AAF in potency [85], whereas chlorination results in a weak carcinogen and iodination leads to a totally inactive compound [98, 100].

The potency of the ring-fluorinated derivatives is in strong contrast to the inactivity of ring-hydroxylated compounds. Ring-hydroxy compounds are in vivo and in vitro metabolites, and 5- and 7-hydroxy are major and 1-, 3- and 9-hydroxy are minor, metabolites [47, 93, 100]. Shielding of the free hydroxy group by forming the methoxy derivative restores some of the activity [36, 100]. The 1- and 3-positional isomers of AAF are also carcinogenic, especially to the mammary gland of female Holtzman rats [37].

Aminoazo Dyes

It has been generally accepted that AB and its acetylated analog are not carcinogenic (fig. 5) [60, 61]. Also, no tumors are induced with N-OH-AB, or *N*-acetoxyacetyl-AB [118]. There is no doubt about the

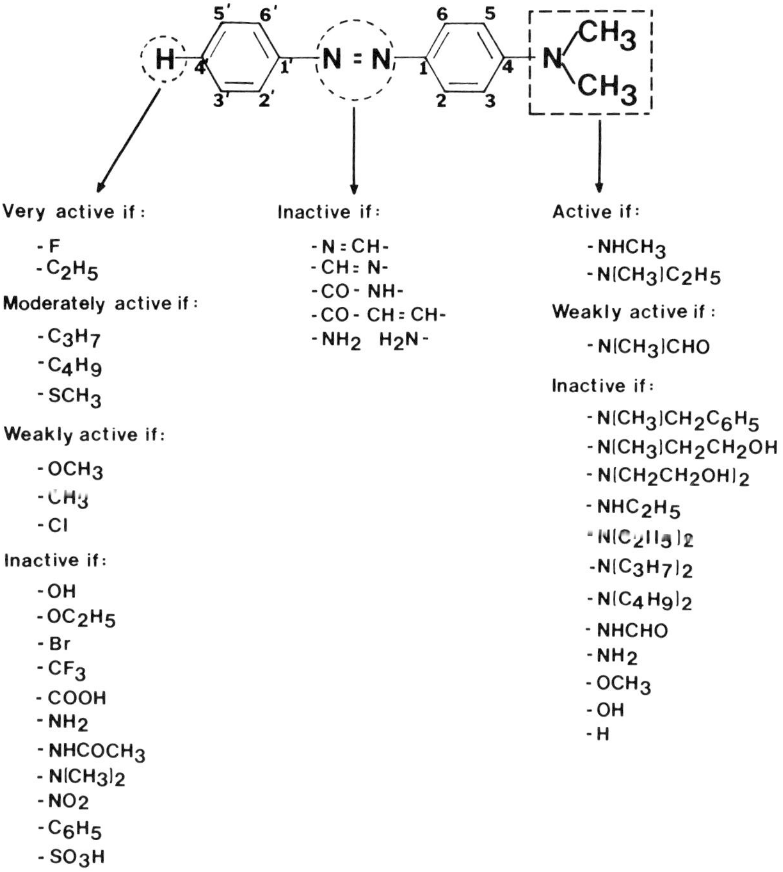

Fig. 5. Structural requirements for carcinogenecity of 4-dimethyl-aminoazobenzenes derivatives. Modified from ref. [2].

carcinogenicity of 2′,3-dimethyl-AB [154]. Five positional isomers of 2′,3-dimethyl-AB were examined in rats and mice, and 5,6′-dimethyl-AB and 4′,6-dimethyl-AB were found to be carcinogenic [20].

DAB or butter yellow is strongly carcinogenic and the azo group is vital for the carcinogenic effects [55]. MAB is, in general, equipotent with DAB [95]. Requirement for at least one methyl group for carcinogenic activity was clearly demonstrated by the finding that *N,N*-diethyl, *N,N*-di-*n*-propyl, *N,N*-di-*n*-butyl, and *N,N*-di-*n*-amyl-AB are not carcinogenic, whereas *N*-ethyl, *N*-methyl-AB and *N*-methyl, *N*-formyl-AB are both potent carcinogens [95, 126].

The two rings in the azo compounds must be coplanar to each other for conjugation across the molecule. Bulky substituents ortho to the azo linkage on DAB, like 2-methyl DAB are less potent, whereas 2-fluoro DAB is almost as active as DAB [95]. The activity of the monofluorinated compound can be explained by the fact that the Van der Waals diameter of fluorine is similar to hydrogen so that a coplanar molecule can still be maintained.

Certain substitutions at the 4′ position, like ethyl, isopropyl, and fluoro groups increase the potency relative to DAB, whereas methyl substitution decreases the potency [17]. Introduction of an amino or dimethylamino at the 4′ position abolishes the carcinogenicity of DAB completely [96]. The latter observation is in strong contrast to carcinogens of the aminobiphenyl and aminofluorene series. Addition of second amino group to the 4′ position of 4-aminobiphenyl results in benzidine, a potent carcinogen [18] and adding another amino group to the extended para position of aminofluorene results in another powerful carcinogen, 2,7-bisaminofluorene [101].

Covalent Binding to Macromolecules

Sites of Covalent Binding

Despite the diverse structures that are found in carcinogenic amines, amides, and aminoazo dyes, there is one feature that is common to all these carcinogens. The final reactive form that interacts with tissue macromolecules is an electrophile, i.e. a species that is deficient in electrons. The reactive electrophile seeks out electron-rich centers in tissue macromolecules, such as the S of methionine and cysteine, C-3 of tyrosine, and N-1 and N-3 of histidine in proteins, C-8, N-3, N-7, and O-6 of guanine and N-1, N-3, and N-7 of adenine in DNA and RNA, all sites of high electron density.

Types of Covalent Binding

Proteins. The first demonstration of covalent binding by chemical carcinogens to tissue macromolecules was accomplished by feeding DAB to rats [86]. Alkaline digestion of livers from rats fed or injected with DAB or MAB resulted in the isolation of 3-methylmercapto-MAB [120]. Attachment of the carcinogen to methionine residues in proteins was substantiated by the isolation of ^{35}S-labeled 3-methylmercapto-

MAB from rats given ^{35}S-methonine and DAB and not from rats given ^{35}S-cysteine and DAB. Further proof for the site of binding of the carcinogen was obtained in vitro by the finding that *N*-benzoyloxy-MAB reacts nonenzymatically with methionine to form a water-soluble dye which decomposes either spontaneously or with alkali to 3-methylmercapto-MAB [71, 110].

Further characterization of the protein-bound dye released by hydrolysis of liver proteins resulted in the isolation of 3-(homocystein-S-yl)-MAB. This adduct presumably arises in vivo through the *S*-demethylation of protein bound 3-(methion-S-yl)-MAB since this same adduct can be formed nonenzymatically by reacting of *N*-benzoyloxy-MAB with homocysteine [67]. Two other dye-bound adducts can be isolated from animals receiving MAB. These have been identified as 3-(3-tyrosyl)-MAB and *N*-(3-tyrosyl)-MAB, formed by alkylation of tyrosine residues in proteins [68]. Even though N-benzoyloxy-MAB also reacts with tryptophan and cysteine in vitro, no polar dyes containing these amino acids have been detected.

Early work with AAF demonstrated that the administration of ^{14}C-AAF or its *N*-hydroxy derivative to rats and certain other species resulted in the incorporation of a small fraction of the administered radioactivity into tissue proteins [87, 123]. N-OH-AF, N-OH-AAF or 2-nitrosofluorene resulted in greater protein-bound radioactivity than the corresponding amide or fluorene. Similarly, other noncarcinogenic metabolites of AAF, like 1- or 7-hydroxy-AAF, also bound as much radioactivity to protein as AAF [137, 140], indicating that binding of carcinogens to protein was probably not a critical step in tumor formation.

Nucleic Acids. Formation of AAF- and N-OH-AAF-bound derivatives of RNA was first demonstrated by Marroquin and Farber [74]. The amount of RNA-bound radioactivity was usually 2–3 times greater with the *N*-oxidized derivatives than with AAF and the specific activity of the RNA was generally two thirds of that observed in liver protein [88]. Administration of the noncarcinogenic metabolite, ^{14}C-1-hydroxy-AAF, resulted in only very low levels of RNA-bound ^{14}C in rat liver. Williard and Irving [143] reported significiant labeling of rat liver DNA after administration of AAF or the *N*-hydroxy derivative, but little binding to DNA after 1-hydroxy-AAF.

The synthetic esters of *N*-hydroxy AAF react readily with guanine residues in nucleosides, nucleotides or nucleic acids and to a lesser extent with adenine residues [66, 83, 85]. In contrast to the attack of direct-alkylating agents at the N-7 position of guanine, esters of *N*-hydroxy AAF react predominantly at the C-8 position of guanine [64, 66]. Thus, the reaction products of *N*-acetoxy AAF with guanosine and deoxyguanosine were identified as *N*-(guanosin-8-yl)-AAF and *N*-(deoxyguanosin-8-yl)-AAF, respectively [66, 83, 85].

Animal studies have indicated that the major AAF adduct in rat liver and mammary gland after administration of the carcinogen is *N*-(deoxyguanosin-8-yl)-AF [1, 9, 63] with minor adducts being *N*-(deoxyguanosin-8-yl)-AAF [64, 65] and 3-(deoxyguanosin-N^2-yl)-AAF [80, 142]. After chronic feeding of AAF to rats, between 80 and 90% of the bound fluorene residues in rat hepatic DNA did not retain the *N*-acetyl group [9, 49, 65, 112]. The relative amount of each fluorene derivative bound to DNA is probably a reflection of the biological half-life of each adduct. After a single dose of AAF, *N*-(deoxyguanosin-8-yl)-AAF has a half-life of only 7 days [65], whereas *N*-(deoxyguanosin-8-yl)-AF persists for several months [9, 112]. The mechanism for formation of deacetylated AAF is suggested by the finding that when the *O*-glucuronide conjugate of *N*-hydroxy-AAF reacts with guanosine, one third of the product is *N*-(guanosin-8-yl)-AAF and about two thirds is the deacetylated product *N*-(guanosin-8-yl)-AF [84].

Accumulation and removal of hepatic DNA adducts after 3 days or 28 days of oral feeding of AAF were examined in male Wistar-Furth rats [111]. DNA adduct formation remained constant during the month of AAF administration. The rate of removal of adduct formed after acute feeding (1–3 days) was faster than after chronic feeding (28 days). Continued AAF administration resulted in a slow accumulation of persistent adducts. Based on the data, two genomic regions were postulated, an area susceptible to fast repair and a region more resistant to the removal of AAF adducts. Overall, the removal of adducts formed during chronic AAF feeding was very efficient since greater than 93% of the adducts were removed by the end of a subsequent month of control diet.

In vitro studies using *N*-benzoyloxy-MAB have demonstrated that after reacting with guanosine or deoxyguanosine, the products formed are *N*-(guanosin-8-yl)-MAB and *N*-(deoxyguanosin-8-yl)-MAB, respectively [70]. Similar results were obtained after the administration of MAB [69]. Tarpley et al. [125] showed that a second major adduct was formed upon reaction of *N*-benzoyloxy-MAB with DNA and this new adduct can also be detected in hepatic DNA of 12-day old mice administered either MAB or DAB. This adduct was identified to be 3-(deoxyguanosin-N^2-yl)-MAB [10].

Single administration of MAB to rats resulted in the detection of only two hepatic MAB-DNA adducts, 3-(deoxyguanosin-N^2-yl)-MAB and *N*-deoxyguanosin-8-yl)-MAB which accounted for 25 and 70%, respectively, of the total MAB bound to DNA at 8 h after administration [127]. After chronic administration of MAB to rats, three DNA adducts were identified, 3-(deoxyguanosin-N^2-yl)-MAB, *N*-(deoxyguanosin-8-yl)-MAB, and 3-(deoxyguanosin-N^6-yl)-MAB, which accounted for 51, 25, and 23%, respectively, of the total adducts. These findings are consistent with the rapid removal of *N*-(deoxyguanosin-8-yl)-MAB and the relative persistence of the 3-(deoxyguanosin-N^2-yl)-MAB and 3-(deoxyguanosin-N^6-yl)-MAB [127].

PHS-dependent metabolism of benzidine in the presence of DNA or RNA results in the covalent binding of activated metabolites to the nucleic acids [51). The incubation of benzidine diimine, a reactive intermediate in the peroxidation of benzidine, with DNA results in an adduct identified to be *N*-(deoxyguanosin-8-yl)-benzidine [153].

Incubation of dog urinary bladder microsomes with N-OH-2-NA, 2-nitrosonaphthalene and 2-amino-1-naphthol resulted in high levels of protein binding for all three derivatives, but only N-OH-2-NA and 2-amino-1-naphthol bound appreciably to DNA. Acid hydrolysis followed by high-performance liquid chromatography analysis of the DNA reacted with PHS and 2-amino-1-naphthol indicated that the major adduct was N^4-(deoxyguanosin-N^2-yl)-2-amino-1,4-naphthoquinoneimine and two minor metabolites tentatively identified as N^4-(deoxyadenosin-N^6-yl)-2-amino-1,4-naphthoquinoneimine and a deoxyguanosin-N^2-yl adduct of a naphthoquinoneimine dimer [152]. These adducts accounted for approximately 20% of the 2-NA bound to urothelial DNA in vivo, suggesting that a minor proportion of the DNA adducts found in vivo may be formed by PHS activation of 2-NA in the target tissue.

Mechanisms of Chemical Carcinogenesis

It is apparent from the previous sections that great progress has been made in identifying the in vivo reactive forms and the nature of the nucleophilic targets of many of the N-containing carcinogens. However, we still do not know unequivocally the nature of the molecular targets critical to the induction of tumors. Lack of knowledge of the chemical nature and biological significance of the critical molecular targets results in only general hypotheses to explain the mechanism(s) of action of chemical carcinogens.

At present, there are at least two proposed mechanisms, one genetic and the other epigenetic, for chemical carcinogenesis. The fundamental premise of the epigenetic hypothesis is that the genetic information of tumor cells need not be altered from that of normal cells of the same organism for tumor formation whereas the genetic hypothesis assumes that the change from normal to tumor cells is dependent on genomic alterations.

Support for epigenetic mechanisms stem mainly from studies with cell culture and cell transplantations. McKinell et al. [78] and Gurdon [33] transplanted nuclei from frog renal carcinomas into enucleated fertilized frog eggs with the subsequent development of apparently normal swimming tadpoles. The potential for differentiation of malignant cells to nonmalignant cells was demonstrated for several types of tumors including cloned teratocarcinoma cells and stem cells from a transplantable squamous cell carcinoma [106]. All these results suggest that induction of tumors by chemicals would be through proliferation of cell lines with altered transcription control [107].

If tumor formation is due to altered genetic information, then such genomic changes in cells would be expected to occur more frequently as a result of direct modification of DNA by carcinogens. Early attempts to correlate carcinogenic and mutational properties of carcinogens met with little success until the realization that metabolic activation is required. A qualitative correlation between mutagenicity and carcinogenicity exists when this consideration is taken into account [76, 77]. This correlation between mutagenicity and carcinogenicity is based on the fact that reactant forms of both mutagens and carcinogens are highly reactive electrophiles. Since these electrophiles also react with RNA and protein, this correlation alone, however, is not sufficient to show that carcinogenesis involves only mutagenic events.

In addition, at least two indirect mechanisms for chemical carcinogens have been proposed. Both mechanisms are likely to operate via the genetic or epigenetic pathways. These two indirect mechanisms are the activation of a latent carcinogenic virus genome and the selection of preneoplastic cells by conditions that favor the multiplication of these cells [92].

Conceivably, any one or a combination of several direct and indirect mechanisms may be involved in the carcinogenic process. Only with progress in cellular and molecular biology will the mechanism of chemical carcinogenesis be understood at the molecular and cellular level.

Concluding Remarks

It has been over 25 years since the *N*-hydroxylation pathway was first demonstrated in vivo. Since that time, *N*-hydroxylation has been established unequivocally as the first critical metabolic step in transforming nitrogen-containing xenobiotics into carcinogens. Even though *N*-oxidation is an essential step in the metabolic activation of some chemical carcinogens, it is by no means a sufficient condition for the initiation of cancer. In many instances, these carcinogens are activated in vivo by multiple and perhaps competing pathways which ultimately result in neoplasia. These multiple pathways probably account for the observed differences in tissue and species susceptibilities to tumor formation.

References

1 Allaben, W.T.; Weis, C.C; Fullerton, N.F.; Beland, F.A.: Formation and persistence of DNA adducts from the carcinogen *N*-hydroxy-2-acetylaminofluorene in rat mammary gland in vivo. Carcinogenesis *4:* 1067–1070 (1983).

2 Arcos, J.C.; Argus, M.F.: Chemical induction of cancer. Structural basis and biological mechanisms, vol. IIB (Academic Press, New York 1974).

3 Bahl, O.P.; Gutmann, H.R.: On the binding of the carcinogen *N*-2-fluorenylacetamide to rat serum albumin in vivo. Biochim. biophys. Acta. *90:* 391–393 (1964).

4 Bartsch, H.; Dworkin, C.; Miller, E.C.; Miller, J.A.: Formation of electrophilic N-acetoxyarylamines in cytosols from rat mammary gland and other tissues by transacetylation from the carcinogen *N*-hydroxy-4-acetylaminobiphenyl. Biochim. biophys. Acta *304:* 42–55 (1973).

5 Bartsch, H.; Dworkin, M.; Miller, J.A.; Miller, E.C.: Electrophilic N-acetoxyaminoarenes derived from carcinogenic *N*-hydroxy-*N*-acetylaminoarenes by enzymatic deacetylation and transacetylation in liver. Biochim. biophys. Acta *286:* 272–298 (1972).

6 Bartsch, H.; Hecker, E.: On the metabolic activation of the carcinogen *N*-hydroxy-*N*-2-acetylaminofluorene. 3. Oxidation with horse radish peroxidase to yield 2-nitrosofluorene and *N*-acetoxy-*N*-2-acetylaminofluorene. Biochim. biophys. Acta *237:* 567–578 (1971).

7 Bartsch, H.; Miller, J.A.; Miller, E.C.: *N*-Acetoxy-*N*-acetylaminoarenes and nitrosoarenes. One electron non-enzymatic and enzymatic oxidation products of various aromatic acethydroxamic acids. Biochim. biophys. Acta *273:* 40–51 (1972).
8 Bartsch, H.; Traut, M.; Hecker, E.: On the metabolic activation of *N*-hydroxy-*N*-2-acetylaminofluorene. II. Simultaneous formation of 2-nitrosofluorene and *N*-acetoxy-N-2-acetylaminofluorene from *N*-hydroxy-*N*-2-acetylaminofluorene via a free radical intermediate. Biochim. biophys. Acta *237:* 556–566 (1971).
9 Beland, F.A.; Dooley, K.L.; Jackson, C.D.: Persistence of DNA adducts in rat liver and kidney after multiple doses of the carcinogen *N*-hydroxy-2-acetylaminofluorene. Cancer Res. *42:* 1348–1354 (1982).
10 Beland, F.A.; Tullis, D.L.; Kadlubar, F.F.; Straub, K.M.; Evans, F.E.: Characterization of DNA adducts of the carcinogen *N*-methyl-4-aminoazobenzene in vitro and in vivo. Chem. Biol. Interact. *31:* 1–18 (1980).
11 Bonser, G.M.; Clayson, D.B.; Juli, J.W.: An experimental inquiry into the cause of industrial bladder cancer. Lancet *ii:* 286–288 (1951).
12 Boyd, J.A.; Eling, T.E.: Metabolism of aromatic amines by prostaglandin H synthase. Environ. Health Perspect. *64:* 45–51 (1985).
13 Boyd, J.A.; Harvans, D.J.; Eling, T.J.: The oxidation of 2-aminofluorene by prostaglandin, endoperoxide synthetase: comparison with other peroxidases. J. biol. Chem. *258:* 8246–8254 (1983).
14 Boyland, E.; Dukes, C.E.; Grove, P.L.: Carcinogenecity of 2-naphthylhydroxylamine and 2-naphthylamine. Br. J. Cancer *17:* 79–84 (1963).
15 Boyland, E.; Manson, D.: The biochemistry of aromatic amines: the metabolism of 2-napthylamine and 2-napthylhydroxylamine derivatives. Biochem. J. *101:* 84–102 (1966).
16 Brill, E.; Radomski, J.: *N*-Hydroxylation of 1-napthylamine in the dog. Life Sci. *6:* 2293–2297 (1967).
17 Brown, E.V.; Hamdan, A.A.: The hepatocarcinogenicity of a series of 4′-alkyl-4-dimethylaminoazobenzene. J. natn. Cancer Inst. *27:* 663–666 (1961).
18 Case, R.A.M.; Hosker, M.E.; McDonald, D.B.; Pearson, J.T.: Tumors of the urinary bladder in workmen engaged in the manufacture and use of certain dyestuff intermediates in the British chemical industry. Br. J. Ind. Med. *11:* 75–104 (1954).
19 Case, R.A.M.; Pearson, J.T.: Tumors of the urinary bladder in workmen engaged in the manufacture and use of certain dyestuff intermediates in the British chemical industry. II. Further consideration of the role of aniline and of the manufacture of auramine and magenta (fuchsine) as possible causative agents. Br. J. Ind. Med. *11:* 213–216 (1954).
20 Crabtree, H.G.: The relative carcinogenic activity of six isomeric aminoazotoluenes. Br. J. Cancer *3:* 387–398 (1949).
21 Cramer, J.W.; Miller, J.A.; Miller, E.C.: N-Hydroxylation: a new metabolic reaction observed in the rat with the carcinogen 2-acetylaminofluorene. J. biol. Chem. *235:* 885–888 (1960).
22 Debaun, J.R.; Miller, E.C.; Miller, J.A.: *N*-Hydroxy-2-acetylaminofluorene sulfotransferase: its probable role in carcinogenesis and in protein-(methion-*S*-yl) binding in rat liver. Cancer Res. *30:* 577–595 (1970).
23 Druckrey, H.: Konstitution und Wirkung cancerogener Substanzen. Arzneimittel-Forsch. *2:* 503–506 (1952).

24 Dyer, H.M.; Kelly, M.G.; O'Gara, R.W.: Lack of carcinogenic activity and metabolic fate of fluorenylacetamides in monkeys. J. natn. Cancer Inst. *36:* 305–322 (1966).
25 Eling, T.; Boyd, J.; Reed, G.; Mason, R.; Sivarajah, K.: Xenobiotic metabolism by prostaglandin endoperoxide synthetase. Drug Metab. Rev. *14:* 1023–1053 (1983).
26 Enomoto, M.; Lotlikar, P.; Miller, J.A.; Miller, E.C.: Urinary metabolites of 2-acetylaminofluorene and related compounds in the rhesus monkey. Cancer Res. *22:* 1336–1342 (1962).
27 Flammang, T.J.; Kadlubar, F.F.: Acetyl coenzyme A-dependent metabolic activation of *N*-hydroxy-3,2′-dimethyl-4-amino biphenyl and several carcinogenic *N*-hydroxylarylamines in relation to tissue and species differences, other acyldonors and arylhydroxamic acid-dependent acyltransferases. Carcinogenesis *7:* 919–926 (1986).
28 Flammang, T.J.; Westra, J.G.; Kadlubar, F.F.; Beland, F.A.: DNA adducts formed from the probable proximate carcinogen, *N*-hydroxy-3-2′-dimethyl-4-aminobiphenyl, by acid catalysis or *S*-acetyl coenzyme A-dependent enzymatic esterification. Carcinogenesis *6:* 251–258 (1985).
29 Floyd, R.A.: Microsome catalyzed conversion of *N*-hydroxy-*N*-acetyl-2-aminofluorene by cumene-hydroperoxide. Life Sci. *18:* 189–195 (1976).
30 Floyd, R.A.: Free radicals in arylamine carcinogenesis; in Pryor, Free radicals in biology, vol. IV, pp. 187–208 (Academic Press, New York 1980).
31 Forrester, A.R.; Oglivy, M.M.; Thomson, R.H.: Mode of action of carcinogenic amines. I. Oxidation of *N*-arylhydroxamic acids. J. chem. Soc. C *8:* 1081–1083 (1970).
32 Grantham, P.H.; Weisburger, E.K.; Weisburger, J.H.: Dehydroxylation and deacetylation of *N*-hydroxy-*N*-2-fluorenylacetamide by rat liver and brain homogenates. Biochim. biophys. Acta *107:* 414–424 (1965).
33 Gurdon, J.B.: The control of gene expression in animal development (Harvard University Press, Cambridge 1974).
34 Gutmann, H.R.; Peters, J.H: Studies on the metabolism of 2-benzoylaminofluorene-9-C^{14} and 2-acetylaminofluorene-9-C^{14} in the rat. Cancer Res. *13:* 415–421 (1953).
35 Gutmann, H.R.; Galitski, S.B.; Foley, W.A.: The conversion of noncarcinogenic aromatic amides to carcinogenic arylhydroxaminic acid by synthetic *N*-hydroxylation. Cancer Res. *27:* 1443–1455 (1967).
36 Gutmann, H.R. Galitski, S.B.; Foley, W.A.: The carcinogenicity of the *O*-methoxy derivatives of *N*-2-fluorenylacetamide and of related compounds in the rat. Cancer Res. *28:* 234–244 (1968).
37 Gutmann, H.R.; Leaf, D.S.; Yost, Y.; Rydell, R.E.; Chen, C.C.: Structure-Activity relationship of N-acetylaryl hydroxylamines in the rat. Cancer Res. *30:* 1485–1498 (1970).
38 Hecker, E.; Traut, M.; Hopp, M.: Über die carcinogene Wirkung von 3-Amino-1,3,5-(10)-oestratrienen und von 2-Nitrosofluoren. Z. Krebsforsch. *71:* 81–88 (1968).
39 Hammons, G.J.; Guengerich, F.P.; Weis, C.C.; Beland, F.A.; Kadlubar, F.F.: Metabolic oxidation of carcinogenic arylamines by rat, dog and human hepatic microsomes and by purified flavincontaining and cytochrome P-450 monooxygenases. Cancer Res. *45:* 3578–3585 (1985).
40 Higginson, J.: Present trends in cancer epidemiology. Can. Cancer Conf. *8:* 40–75 (1969).

41 Heubner, W.: Studien über Methämoglobinbildung. Arch. exp. Path. Pharmak. *72:* 241–281 (1915).
42 Hueper, W.C.; Wiley, F.H.; Wolfe, H.D.: Experimental production of bladder tumor in dogs by administration of beta-napthylamine. J. ind. hyg. Toxicol. *20:* 46–84 (1938).
43 IARC: Evaluation of the carcinogenic risks of chemical to man; Vol. 4, pp. 27–124 (International Agency of Research on Cancer, Lyon 1974).
44 Irving, C.C.: *N*-Hydroxylation of 2-acetylaminofluorene in the rabbit. Cancer Res. *22:* 867–873 (1962).
45 Irving, C.C.: Metabolic activation of *N*-hydroxy compounds by conjugation. Xenobiotica *1:* 387–398 (1971).
46 Irving, C.C.: Metabolic activation of aromatic amines; in Clark, Shaw, Carcinogens: identification and mechanism of action, pp. 211–227 (Raven Press, New York 1979).
47 Irving, C.C.; Gutmann, H.R.; Larson, D.M.: Evaluation of the carcinogenicity of aminofluorenols by implantation into the bladder of the mouse. Cancer Res. *23:* 1782–1790 (1963).
48 Irving, C.C.; Janss, D.H.; Russell, L.T.: Lack of *N*-hydroxy-2-acetylaminofluorene sulfotransferase activity in the mammary gland and Zymbal's gland of the rat. Cancer Res. *31:* 387–391 (1971).
49 Irving, C.C.; Veazey, R.A.: Persistent binding of 2-acetylaminofluorene to rat liver DNA in vivo and consideration of the mechanism of binding of *N*-hydroxy-2-acetylaminofluorene to rat liver nucleic acids. Cancer Res. *29:* 1799–1804 (1969).
50 Irving, C.C.; Veazey, R.A.: Differences in the binding of 2-acetylaminofluorene and its *N*-hydroxy metabolite to liver nucleic acids of male and female rats. Cancer Res. *31:* 19–22 (1971).
51 Kadlubar, F.F.; Frederick C.B.; Weis, C.C.; Zenser, T.V.: Prostaglandin endoperoxide synthetase-mediated metabolism of carcinogenic aromatic amines and their binding to DNA and protein. Biochem. biophys. Res. Commun. *108:* 253–258 (1982).
52 Kadlubar, F.F.; Miller, J.A.; Miller, E.C.: Hepatic metabolism of *N*-hydroxy-*N*-methyl-4-amino-benzene and other N-hydroxyarylamines to reactive sulfuric acids esters. Cancer Res. *36:* 2350–2359 (1976).
53 Kadlubar, F.F.; Miller, J.A.; Miller E.C.: Hepatic microsomal *N*-glucuronidation and nucleic acid binding of *N*-hydroxyarylamines in relation to urinary bladder carcinogenesis. Cancer Res. *37:* 805–814 (1977).
54 Kadlubar, F.F.; Unruh, L.E.; Flammang, T.J.; Sparks, D.; Mitchum, R.K.; Mulder, G.J.: Alterations of urinary levels of the carcinogen, *N*-hydroxy-2-naphthylamine, and its *N*-glucuronide in the rat by control of urinary pH, inhibition of metabolic sulfation, and changes in biliary excretion. Chem.-Biol. Interact. *33:* 129–147 (1981).
55 Kenseler, C.J.; Dexter, S.D.; Rhoads, C.P.: The inhibition of a diphosphopyridine nucleotide system by split products of dimethylaminoazobenzene. Cancer Res. *2:* 1–10 (1942).
56 King, C.M.: Mechanism of action, tissue distribution, and inhibition of arylhydroxymic acid acyltransferase. Cancer Res. *34:* 1503–1515 (1974).
57 King, C.M.; Phillips, B.: Enzyme catalyzed reactions of the carcinogen *N*-hydroxy-2-fluorenylacetamide with nucleic acids. Science *159:* 1351–1353 (1968).

58 King, C.M.; Traub, N.R.; Lortz, Z.M.; Thissen, M.R.: Metabolic activation of arylhydroxamic acids by N-O-acyltransferase of rat mammary gland. Cancer Res. *39:* 3369–3372 (1979).
59 Kinosita, R.: Researches on the carcinogenesis of the various chemical substances. Gann. *30:* 423–426 (1936).
60 Kirby, A.H.M.: Attempts to induce stomach tumors. III. The effects of (a) a residue of cholesterol heated to 300 °C and (b) 3,5-cholestadiene. Cancer Res. *4:* 94–97 (1944).
61 Kirby, A.H.M.: Studies on carcinogenesis with azo compounds. III. The action of (a) four azo compounds in Wistar rats fed restricted diet; (b) *N,N*-dimethyl-*p*-aminoazobenzene in mice. Cancer Res. *7:* 333–334 (1947).
62 Kriek, E.: On the interaction of N-2-fluorenylhydroxylamine with nucleic acid in vitro. Biochem. biophys. Res. Commun. *20:* 793–799 (1965).
63 Kriek, E.: Difference in binding of 2-acetylaminofluorene to rat liver deoxyribonucleic acid and ribosomal ribonucleic acid in vivo. Biochim. biophys. Acta *161:* 273–275 (1968).
64 Kriek, E.: On the mechanism of action of carcinogenic aromatic amines. 1. Binding of 2-acetylaminofluorene and *N*-hydroxy-2-acetylaminofluorene to rat liver nucleic acids in vivo. Chem.-Biol. Interact. *1:* 1–17 (1969).
65 Kriek, E.: Persistent binding of a new reaction product of the carcinogen *N*-hydroxy-*N*-2-acetylaminofluorene with guanine in rat liver DNA in vivo. Cancer Res. *32:* 2042–2048 (1972).
66 Kriek, E.; Miller, J.A.; Juhl, U.; Miller, E.C.: 8-(*N*-2-fluorenylacetamide-guanosine, an arylamidation reaction product of guanosine and the carcinogen *N*-acetoxy-*N*-fluorenylacetamide in neutral solution. Biochemistry *6:* 177–182 (1967).
67 Lin, J.K.; Miller, J.A.; Miller, E.C.: Studies on structure of polar dyes derived from the liver proteins of rats fed *N*-methyl-4-aminoazobenzene. II. Identity of synthetic 3-*C*-homocystein-*S*-yl)-*N*-methyl-4-aminoazobenzene with the major polar dye p2b. Biochemistry *7:* 1889–1895 (1968).
68 Lin, J.K.; Miller, J.A.; Miller, E.C.; Studies on structure of polar dyes derived from the liver proteins of rats fed *N*-methyl-4-aminoazobenzene. III. Tyrosine and homocysteine sulfoxide polar dyes. Biochemistry *8:* 1573–1582 (1969).
69 Lin, J.K.; Miller, J.A.; Miller, E.C.: Structure of hepatic nucleic acid-bound dyes in rats given the carcinogen *N*-methyl-4-aminoazobenzene. Cancer Res. *35:* 844–850 (1975).
70 Lin, J.K.; Schmall, B.; Sharpe, I.D.; Miura, I.; Miller, J.A.; Miller, E.C.: N-Substitution of carbon 8 in guanosine and deoxyguanosine by the carcinogen *N*-benzoyloxy-*N*-methyl-4-aminoazobenzene in vitro. Cancer Res. *35:* 832–843 (1975).
71 Lotlikar, P.D.; Scribner, J.D.; Miller, J.A.; Miller, E.C.: Reaction of esters of aromatic *N*-hydroxyamines and amides with methionine in vitro: a model for in vivo binding of amine carcinogens to proteins. Life Sci. *5:* 1263–1269 (1966).
72 Magee, P.N.; Montesano, R.; Preussmann, R.: N-Nitroso compounds and related carcinogens; in Searle, Chemical carcinogens, pp. 491–625. (Am. Chemical Society, Washington 1976).
73 Marnett, L.J.; Wlodawer, P.; Samuelsson, B.: Co-oxygenation of organic substrates by the prostaglandin synthetase of sheep vesicular gland. J. biol. Chem. *250:* 8510–8517 (1975).

74 Marroquin, R.F.; Farber, E.: The binding of 2-acetylaminofluorene to rat liver nucleic acids in vivo. Cancer Res. *25:* 1262–1269 (1965).
75 Matthews, J.J.; Walpole, A.L.: Tumors of the liver and kidney induced in Wistar rats with 4′-fluoro-4-aminodiphenyl. Br. J. Cancer *12:* 234–241 (1958).
76 McCann, J.; Ames, B.N.: Detection of carcinogens as mutagens in the Salmonella microsome test: assay of 300 chemicals. Discussion. Proc. natn. Acad. Sci. USA *73:* 950–954 (1976).
77 McCann, J.; Choi, E.; Yamasaki, E.; Ames, B.N.: The detection of carcinogens as mutagens in the Salmonella microsome test: assay of 300 chemicals. Proc. natn. Acad. Sci. USA *72:* 5135–5139 (1975).
78 McKinnel, R.G.; Deggins, B.A.; Labat, D.D.: Transplantation of pluripotential nuclei from triploid frog tumors. Science *165:* 394–396 (1969).
79 McManus, M.E.; Minichin, R.F.; Sanderson, N.; Wirth, P.J.; Thorgeirsson, S.S.: Kinetics of *N*- and *C*-hydroxylations of 2-acetylaminofluorene in male Sprague-Dawley rat liver microsomes: implications for carcinogenesis. Cancer Res. *43:* 3720–3724 (1983).
80 Meerman, J.H.N.; Beland, F.A.; Mulder, G.J.: Role of sulfation in the formation of DNA adducts from *N*-hydroxy-2-acetylaminofluorene in rat liver in vivo. Inhibition of *N*-acetylated aminofluorene adduct formation by pentachlorophenol. Carcinogenesis *2:* 413–416 (1981).
81 Meerman, J.H.N.; Mulder, G.J.: Prevention of the hepatotoxic action of *N*-hydroxy-2-acetylaminofluorene in the rat by inhibition of N-O-sulfation by pentachlorophenol. Life Sci. *28:* 2361–2365 (1981).
82 Miller, E.C.: Some current perspectives on chemical carcinogenesis in humans and experimental animals: presidential address. Cancer Res. *38:* 1469–1479 (1978).
83 Miller, E.C.; Juhl, U.; Miller, J.A.: Nucleic acid guanine: reaction with the carcinogen *N*-acetoxy-2-acetylaminofluorene. Science *153:* 1125–1127 (1966).
84 Miller, E.C.; Lotlikar, P.D.; Miller, J.A.; Butler, B.W.; Irving, C.C.; Hill, J.J.: Reactions in vivo of some tissue nucleophiles with the glucuronide of *N*-hydroxy-2-acetylaminofluorene. Molec. Pharmacol. *4:* 147–154 (1968).
85 Miller, E.C.; Lotlikar, P.D.; Pitot, H.C.; Fletcher, T.L.; Miller, J.A.: *N*-hydroxy metabolite of 2-acetylaminophenanthrene and 7-fluoro-2-acetylaminofluorene as proximate carcinogens in the rats. Cancer Res. *26:* 2239–2247 (1966).
86 Miller, E.C.; Miller, J.A.: The presence and significance of bound aminoazo dyes in the livers of rats fed *p*-dimethylaminoazobenzene. Cancer Res. *7:* 468–480 (1947).
87 Miller, E.C.; Miller, J.A.: Biochemical investigations on hepatic carcinogens. J. natn. Cancer Inst. *15:* 1571–1590 (1955).
88 Miller, E.C.; Miller, J.A.: Mechanisms of chemical carcinogenesis: nature of proximate carcinogens and interactions with macromolecules. Pharmacol. Rev. *18:* 805–838 (1966).
89 Miller, E.C.; Miller, J.A.; Enomoto, M.: The comparative carcinogenicities of 2-acetylaminofluorene and its *N*-hydroxy metabolite in mice, hamsters, and guinea pigs. Cancer Res. *24:* 2018–2031 (1964).
90 Miller, E.C.; Miller, J.A.; Hartmann, H.A.: *N*-Hydroxy-2-acetylaminofluorene: a metabolite of 2-acetylaminofluorene with increased carcinogenic activity in the rat. Cancer Res. *21:* 815–824 (1961).
91 Miller, E.C.; Sandin, R.B.; Miller, J.A.; Rusch, H.P.: The carcinogenicity of com-

pounds related to 2-acetylaminofluorene. IV. Aminobiphenyl and benzidine derivatives. Cancer Res. *16:* 525–534 (1956).
92 Miller, J.A.: Carcinogenesis by chemicals: an overview – G.H.A. Clowes memorial lecture. Cancer Res. *30:* 559–576 (1970).
93 Miller, J.A.; Cramer, J.W.; Miller, E.C.: The *N*- and ring-hydroxylation of 2-acetylaminofluorene during carcinogenesis in the rat. Cancer Res. *20:* 950–962 (1960).
94 Miller, J.A.; Enomoto, M.; Miller, E.C.: The carcinogenecity of small amounts of *N*-hydroxy-2-acetylaminofluorene and its cupric chelate in the rat. Cancer Res. *22:* 1381–1388 (1962).
95 Miller, J.A.; Miller, E.C.: The carcinogenic aminoazo dyes. Adv. Cancer Res. *1:* 339–396 (1953).
96 Miller, J.A.; Miller, E.C.; Finger, G.C.: Further studies on the carcinogenecity of dyes related to 4-dimethylaminoazobenzene. The requirement for an unsubstituted 2-position. Cancer Res. *17:* 387–398 (1957).
97 Mirvish, S.S.: Formation of *N*-nitroso compounds: chemistry, kinetics, and in vivo occurrence. Toxicol. Appl. Pharmacol. *31:* 325–351 (1975).
98 Morris, H.P.: Hepatocarcinogenesis by 2-acetylaminofluorene and related compounds including comments on dietary and other influences. J. natn. Cancer Inst. *15:* 1535–1545 (1955).
99 Morris, H.P.; Dubnik C.S.; Johnson, J.M.: Studies on the carcinogenic action in the rat of 2-nitro, 2-amino, 2-acetylamino, and 2-diacetylaminofluorene after injection and after painting. J. natn. Cancer Inst. *10:* 1201–1213 (1950).
100 Morris, H.P.; Velat, C.A.; Wagner, B.P.; Dahlgard, M.; Ray, F.E.: Studies of carcinogenecity in the rat of derivatives of aromatic amines related to *N*-2-fluorenylacetamide. J. natn. Cancer Inst. *24:* 149–180 (1960).
101 Morris, H.P.: Wagner, B.P.; Ray, F.E.; Snell, K.C.; Stewart, H.L.: Comparative study of cancer and other lesions fed *N,N'*-2,7-fluorenylenebisacetamide of N-2-fluorenylacetamide. Natn. Cancer Inst. Mongr. *5:* 1–30 (1961).
102 Morton, K.C.; King, C.M.; Vaught, J.B.; Wang, C.Y.; Lee, M.S.; Marnett, L.J.: Prostaglandin H synthase-mediated reaction of carcinogenic arylamines with t-RNA and homopolyribonucleotides. Biochem. biophys. Res. Commun. *111:* 96–103 (1983).
103 Mulder, G.J.; Scholtens, E.: Phenol sulfotransferase and uridine diphosphate glucuronosyltransferase from rat liver in vivo and in vitro: 2,6-Dichloro-4-nitrophenol as selective inhibitor of sulphation. Biochem. J. *165:* 553–559 (1977).
104 Ogleby, L.A.; Flammang, T.J.; Tullis, D.L.; Kadlubar, F.F.: Rapid absorption, distribution, and excretion of carcinogenic *N*-hydroxy-arylamines after direct urethral instillation into the rat bladder. Carcinogenesis *2:* 15–20 (1981).
105 Peters J.H.; Gutmann, H.R.: The acetylation of 2-acetylaminofluorene and the deacetylation and concurrent reacetylation of 2-acetylaminofluorene by rat liver slices. J. biol. Chem. *216:* 713–728 (1955).
106 Pierce, G.B.: Differentiation of normal and malignant cells. Fed. Proc. *29:* 1248–1254 (1970).
107 Pitot, H.C.; Heidelberger, C.: Metabolic regulatory circuits and carcinogenesis. Cancer Res. *23:* 1694–1700 (1963).
108 Pohl, R.J.; Fouts, J.R.: Cytochrome P-450-dependent xenobiotic metabolizing activity in Zymbal's gland, a specialized sebaceous gland of rodents. Cancer Res. *43:* 3660–3662 (1983).

109 Poirier, L.A.; Miller, J.A.; Miller, E.C.: The *N*- and ring-hydroxylation of 2-acetylaminofluorene and the failure to detect *N*-acetylation of 2-aminofluorene in the dog. Cancer Res. *23:* 790–800 (1963).

110 Poirier, L.A.; Miller, J.A.; Miller, E.C.; Sato, R.: *N*-benzoyloxy-*N*-methyl-4-aminoazobenzene: its carcinogenic activity in the rat and its reactions with proteins and nucleic acids and their constitutents in vitro. Cancer Res. *27:* 1600–1613 (1967).

111 Poirier, M.C.; Hunt, J.M.; True, B.A.; Laishes, B.A.; Young, J.F.; Beland, F.A.: DNA adduct formation, removal and persistence in rat liver during one month of feeding 2-acetylaminofluorene. Carcinogenesis *5:* 1591–1596 (1984).

112 Poirier, M.C.; True, B.; Laishes, B.: Formation and removal of (guan-8-yl)-DNA-2-acetylaminofluorene adducts in liver and kidney of male rats given dietary 2-acetylaminofluorene. Cancer Res. *42:* 1317–1321 (1982).

113 Radomski, J.L.: The primary aromatic amines: their biological properties and structure activity relationships. Annu. Rev. Pharmacol. Toxicol. *18:* 129–157 (1979).

114 Radomski, J.L.; Brill, E.: Bladder cancer induction by aromatic amines: role of *N*-hydroxy metabolite. Science *167:* 992–993 (1970).

115 Radomski, J.L.; Deichmann, W.B.; Altman, N.H.; Radomski, T.: Failure of pure 1-naphthylamine to induce bladder tumors in dogs. Cancer Res. *40:* 3537–3539 (1980).

116 Radomski, J.L.; Hearn, W.L.; Radomski, T.; Moreno, H.; Scott, W.E.: Isolation of the glucuronic acid conjugate of *N*-hydroxy-4-aminobiphenyl from dog urine and its mutagenic activity. Cancer Res. *37:* 1757–1762 (1977).

117 Rehn, L.: Blasengeschwülste bei Fuchsin-Arbeitern. Arch. klin. Chir. *50:* 588–600 (1895).

118 Sato, R.; Poirier, L.A.; Miller, J.A.; Miller, E.C.: Studies on the *N*-hydroxylation and carcinogenicity of 4-aminoazobenzene and related compounds. Cancer Res. *26:* 1678–1687 (1966).

119 Schinz, H.R.; Fritz-Niggli, H.; Campbell, T.W.; Schmid, H.: Krebsbildung durch Aminofluorene und verwandte Körper. Oncologia *8:* 233–245 (1955).

120 Schribner, J.D.; Miller, J.A.; Miller, E.C.: 3-Methyl-mercapto-*N*-methyl-4-aminoazobenzene: an alkaline degradation product of a labile protein-bound dye in the livers of rats fed *N,N*-dimethyl-4-aminoazobenzene. Biochem. biophys. Res. Commun. *20:* 560–565 (1965).

121 Schribner, J.D.; Miller, J.A.; Miller, E.C.: Nucleophilic substitution on carcinogenic *N*-acetoxy-*N*-arylacetamide. Cancer Res. *30:* 1570–1579 (1970).

122 Schut, H.A.; Castonguay, A.: Metabolism of carcinogenic amino derivatives in various species and DNA alkylation by their metabolites. Drug Metab. Rev. *15:* 753–839 (1984).

123 Seal, U.S.; Gutmann, H.R.: On the mechanism of protein binding of the carcinogen *N*-fluorenylacetamide by rat liver. Ion exchange chromatography of the soluble fraction. Archs. Biochem. Biophys. *95:* 1–11 (1961).

124 Sivarajah, K.; Lasker, J.M., Eling, T.E.; Aboudonia, M.B.: Metabolism of *N*-alkyl compounds during the biosynthesis of prostaglandins. Molec. Pharmacol. *21:* 133–141 (1982).

125 Tarpley, W.G.; Miller, J.A.; Miller, E.C.: Adducts from the reaction of *N*-benzoyloxy-*N*-methyl-4-aminoazobenzene with deoxyguanosine or DNA in vitro and from hepatic DNA of mice treated with *N*-methyl- or *N,N*-dimethyl-4-aminoazobenzene. Cancer Res. *40:* 2493–2499 (1980).

126 Terayama, H.: Aminoazo carcinogenesis-methods and biochemical problems. Methods Cancer Res. *1:* 399–449 (1967).
127 Tillis, D.L.; Straub, K.M.; Kadlubar, F.F.: A comparison of the carcinogen-DNA adducts formed in rat liver in vivo after administration of single or multiple doses of *N*-methyl-4-aminoazobenzene. Chem.-Biol. Interact *38:* 15–27 (1981).
128 Ts'o, P.O.P.; Caspary, W.J.; Lorentzen, R.J.: The involvement of free radicals in chemical carcinogenesis; in Pryor, Free radicals in biology, vol. III, pp. 251–303 (Academic Press, New York 1977).
129 Veys, C.A.: Two epidemiological inquiries into the incidence of bladder tumors in industrial workers. J. natn. Cancer Inst. *43:* 219–226 (1969).
130 Walpole, A.L.; Williams, M.H.C.: Aromatic amines as carcinogens in industry. Br. med. Bull. *14:* 141–145 (1958).
131 Walpole, A.L.; Williams, M.H.C.; Roberts, D.C.: The carcinogenic action of 4-aminodiphenyl and 3,2′-dimethyl-4-amino diphenyl. Br. J. Ind. Med. *9:* 255–263 (1952).
132 Weisburger, J.H.: The enzymatic deacetylation of the carcinogen 2-acetylaminofluorene and related compounds. Biochim. biophys. Acta *16:* 382–390 (1955).
133 Weisburger, J.H.; Grantham, P.H.; Vanhorn, E.; Steigbigel, N.H.; Rall, D.P.; Weisburger E.K.: Activation and detoxification of *N*-2-fluorenylacetamide in man. Cancer Res. *24:* 475–479 (1964).
134 Weisburger, J.H.; Grantham, P.H.; Weisburger, E.K.: The metabolism of *N*-2-fluorenylacetamide in the cat: evidence for glucuronic acid conjugates. Biochem. Pharmacol. *13:* 469–475 (1964).
135 Weisburger, J.H.; Grantham, P.H.; Weisburger, E.K.: The metabolism of *N*-2-fluorenylacetamide in the hamster. Toxicol. appl. Pharmacol. *6:* 427–433 (1964).
136 Weisburger, J.H.; Weisburger, E.K.: Biochemical formation and pharmacological, toxicological and pathological properties of hydroxylamines and hydroxamic acids. Pharmacol. Rev. *25:* 1–66 (1973).
137 Weisburger, J.H.; Weisburger, E.K.; Grantham, P.H.; Morris, H.P.: The biosynthetic preparation of C14-labeled *N*-(1-and 3-hydroxy-2-fluorenyl) acetamides, and their metabolism in rats and guinea pigs. J. natn. Cancer Inst. *22:* 825–844 (1959).
138 Weisburger, J.H.; Weisburger, E.K.; Morris, H.P.: The excretion of radioactivity during a four-day period following the feeding of carbon 14-labeled 2-acetylaminofluorene to rats. J. natn. Cancer Inst. *11:* 797–803 (1951).
139 Weisburger, J.H.; Weisburger, E.K.; Morris, H.P.: Analogs of the carcinogen 2-acetylaminofluorene: the isomeric 4-acetylaminofluorene. J. Am. chem. Soc. *74:* 4540–4543 (1952).
140 Weisburger, J.H.; Weisburger, E.K.; Morris, H.P.: The biosynthesis of *N*-(7-hydroxy-2-fluoren-9-C14-yl) acetamide from *N*-2-fluoren-9-C14-yl-acetamide, and its metabolism in the rat and in the guinea pig. Archs. Biochem. Biophys. *80:* 187–198 (1959).
141 Weisburger, J.H.; Yamamoto, R.S.; Williams, G.M.; Grantham, P.H.; Matsushima, T.; Weisburger, E.K.: On the sulfate ester of *N*-hydroxy-*N*-2-fluorenylacetamide as a key ultimate hepatocarcinogen in the rat. Cancer Res. *32:* 491–500 (1972).
142 Westra, J.G.; Kriek, E.; Hittenhausen, H.: Identification of the persistently bound form of the carcinogen *N*-acetyl-2-aminofluorene to rat liver DNA in vivo. Chem.-Biol. Interact. *15:* 149–164 (1976).
143 Williard, R.F.; Irving, C.C.: On the binding of the carcinogen 2-acetylaminofluorene to rat liver RNA and DNA. Fed. Proc. *23:* 164 (1964).

144 Wilson, R.H.; DeEds, F.; Cox, A.J., Jr.: The toxicity and carcinogenic activity of 2-acetylaminofluorene. Cancer Res. *1:* 595–608 (1941).
145 Wilson, R.H.; DeEds, F.; Cox, A.J.: The carcinogenic activity of 2-acetylaminofluorene. II. Effects of concentration and duration of exposure. Cancer Res. *7:* 444–449 (1947).
146 Wilson, R.H.; DeEds, F.; Cox, A.J.: The carcinogenic activity of 2-acetylaminofluorene. III. Manner of administration, age of animals and types of diet. Cancer Res. *7:* 450–452 (1947).
147 Wilson, R.H.; DeEds, F.; Cox, A.J.: The carcinogenic activity of 2-acetylaminofluorene. IV. Action of related compounds. Cancer Res. *7:* 453–458 (1947).
148 Wise, R.W.; Lakshmi, V.M.; Zenser, T.V.; Davis, B.B.: Peroxide-mediated binding of aromatic amine carcinogens to tissue DNA. Fed. Proc. *45:* 276 (1986).
149 Wise, R.W.; Zenser, T.V.; Kadlubar, F.F.; Davis, B.B.: Metabolic activation of carcinogenic aromatic amines by dog bladder and kidney prostaglandin H synthase. Cancer Res. *44:* 1893–1897 (1984).
150 Wise, R.W.; Zenser, T.V.; Rice, J.R.; Davis, B.B.: Peroxidative metabolism of benzidine by intact tissues: a prostaglandin H synthase-mediated process. Carcinogenesis *7:* 111–115 (1986).
151 Wynder, E.L.; Goldsmith, R., Jr.: The epidemiology of bladder cancer: a second look. Cancer *40:* 1246–1268 (1977).
152 Yamazoe, Y.; Miller, D.W.; Weis, C.C.; Dooley, K.L.; Zenser, T.V.; Beland, F.A.; Kadlubar, F.F.: DNA adducts formed by ring oxidation of the carcinogen 2-naphthylamine with prostaglandin H synthase in vitro and in the dog urothelium in vivo. Carcinogenesis *6:* 1379–1387 (1985).
153 Yamazoe, Y.; Roth, R.W.; Kadlubar, F.F.: Reactivity of benzidine diimine with DNA to form *N*-(deoxyguanosin-8-yl)-benzidine. Carcinogenesis *7:* 179–182 (1986).
154 Yoshida, T.: Über die serienweise Verfolgung der Veränderungen der Leber bei experimenteller Hepatomerzeugung durch *O*-Aminoazotoluol. Trans. Jap. Pathol. Soc. *23:* 636–638 (1933).
155 Young, J.F.; Kadlubar, F.F.: A pharmacokinetic model to predict exposure of the bladder epithelium to urinary *N*-hydroxylamine carcinogens as a function of urine pH, voiding interval, and resorption. Drug Metab. Disposit *10:* 641–644 (1982).
156 Zenser, T.V.; Mattammal, M.B.; Armbrecht, H.J.; Davis, B.B.: Benzidine binding to nucleic acids mediated by the peroxidative activity of prostaglandin endoperoxidase synthase. Cancer Res. *40:* 2839–2845 (1980).
157 Zenser, T.V.; Mattammal, M.B.; Davis, B.B.: Co-oxidation of benzidine by renal medullary prostaglandin cyclooxygenase. J. Pharmac. exp. Ther. *211:* 460–464 (1979).

Check Y. Quon, PhD, Clinical Pharmacology and Drug Metabolism, Research and Development, Du Pont Critical Care, Waukegan, Il 60085 (USA)

Prog. bas. clin. Pharmac., vol. 1, pp. 161–183 (Karger, Basel 1988)

VI. Nitrogen Oxidation in Toxicity[1]

R.A. Prough, J.M. Erikson, P. Wiebkin

Department of Biochemistry, The University of Louisville School of Medicine, Louisville, Ky., USA

I. Chemical Reactivity of Oxidized Compounds

While many nitrogenous compounds can react directly to exert certain physiologic and/or toxic effects by competing for the active sites of enzymes or receptors [Gilman et al., 1985], a number of their oxidized products are chemically unstable and, under certain conditions, can react with cellular nucleophiles. This reactivity can lead to toxic responses or can be directed by certain detoxification processes to go undetected. These concepts of metabolic activation and detoxification have been firmly established by the pioneering work of Boyland [1950], Magee and Barnes [1967], the Millers [1966, 1978], and Williams [1959]. The purpose of this chapter is to describe the chemical reactivity of the various nitrogen oxidation products formed by the enzyme systems described in chapters II–IV and encapsulate the evidence that such chemicals can lead to toxic phenomena in biological systems. A distinction must be made between reactions in which the nitrogen atom is oxidized [i.e. electrons are removed as described by Guengerich et al., [1982] or reactions where the nitrogen is oxygenated [i.e. active oxygen forms a coordinate covalent bond with the nitrogen as described by Ziegler, 1980]. Therefore, in this chapter the authors will distinguish between *N*-oxidation and *N*-oxygenation reactions based on their apparent chemical differences.

[1] Portions of the work presented were supported by American Cancer Society Grant BC-336 and Robert A. Welch Foundation Grant I-616.

A. Amine Oxides

Most amine oxides are solids, but due to their polarity, are very hygroscopic [Smith, 1966]. The polarity of amine oxides makes them very water-soluble and consequently poorly soluble in hydrocarbons or other nonpolar solvents. The N-O functional group can be protonated; the pK_a of these compounds ranges from 4.2 to 5.1.

Chemically, these compounds are reasonably stable and not highly susceptible to nucleophilic attack [Smith, 1966]. They are thermally labile and eliminate an N-alkyl group to form a hydroxylamine and an aldehyde [Bamberger and Leyden, 1901; Dunstan and Goulding, 1899]. Mitchell and Ziegler [1969] have provided a general method to measure *N*-methylamine oxides utilizing the susceptibility of these compounds to be degraded to formaldehyde. In addition, the decomposition of amine oxides by acylating agents (the Polonovski reaction) can also be catalyzed by ferric iron, possibly via a one-electron redox step, to yield the corresponding dealkylated amine and an aldehyde [Cymerman-Craig et al., 1961]. This process apparently involves rearrangement by migration of the oxygen from nitrogen to carbon centers. Finally, amine oxides can be reduced facilely back to the primary amine by a number of reducing agents, such as tin and HCl [Gutman, 1922], hydrogen/Pd [Bodendorf and Blinder, 1954], and reduced flavin [Kato et al., 1976].

B. Hydroxylamines

Hydroxylamines are weakly acidic (pK_a 3–5) and can form salts upon protonation with many acids [Smith, 1966]. Deprotonation at high pH causes the compounds to become easily air oxidized. They are nucleophilic due to the presence of the α-heteroatom and can react with carbonyl groups to form stable oximes and nitrones. In addition, mono- and disubstituted hydroxylamines are easily oxidized to form either *C*-nitroso-, nitrone, or nitroxyl radical species depending on the substituents [Bamberger, 1894]. When monosubstituted hydroxylamines are slowly oxidized to form *C*-nitroso-derivatives, the *C*-nitroso- and hydroxylamine derivatives can condense to form azoxy compounds [Smith, 1966]. Four-electron oxidation reactions of monosubstituted hydroxylamines lead to the formation of the nitro-derivative [Bamberger, 1894]. Disubstituted hydroxylamines can be oxidized in a one-electron step to give nitroxyl radicals and these radicals are very stable if the compound has no α-hydrogens [Sulkowski and Childress, 1963]. If the disubstituted hydroxylamine contains an α-hydrogen, a two-electron oxidation can result in formation of nitrones with the nitroxyl radical serving as a transient intermediate [Banfield and Kenyon, 1926].

Aromatic hydroxylamines behave differently in an acidic medium from other hydroxylamines [Bamberger, 1894]. Instead of forming salts, phenylhydroxylamines form significant amounts of hydroquinones and phenols. Booth and Boyland [1964] noted that aromatic hydroxylamines rearranged to form *o*-aminophenols. It has been proposed that such rearrangements involved protonation of the hydroxylamine functional group and elimination of water to form a resonance-stabilized nitrenium cation. This species can be attacked by water or cellular nucleophiles to form phenols or covalently bound adducts of thiols, DNA or other cellular nucleophiles (fig. 1). The formation of such reactive nitrenium cations may account for the deleterious effects that arylamines have in biological systems.

C. N-*Hydroxyamides*

The principal reaction of substituted *N*-hydroxyamides or hydroxamic acids (acyl derivatives of *N*-hydroxyamines) is to rearrange to form phenols or covalent adducts with nucleophiles like those seen for aromatic hydroxylamines [Booth and Boyland, 1964; Heller et all., 1951]. With hydroxamic derivatives, the rearrangement of an oxygen function from a nitrogen to a carbon center is similar to that proposed by Boyland for aromatic hydroxylamines and is known as the Lossen rearrangement [Smith, 1966]. A nitrenium cation similar to that envisioned for aromatic hydroxylamines has been suggested for *N*-hydroxyamides.

D. N-*Nitrosamines*

Chemically, the disubstituted *N*-nitrosamines are very stable, slightly acidic, and very soluble in water [Smith, 1966]. In acid, they undergo a reversible reaction to form an amine and nitrous acid. The reverse reaction also leads to the formation of *N*-nitrosamines and probably is responsible for the widespread environmental distribution of these compounds. In base, these compounds form salts. The *N*-nitrosamines can be oxidized to form nitramines and can be reduced to diazenium salts, hydrazines and the parent amine. The different products noted are the result of successive stages of reduction. The reactivity of hydrazines and diazenes is discussed in section II.E. As will be seen later, the reactivity of *N*-nitrosamines most likely involves oxidation of a carbon center, not a nitrogen center.

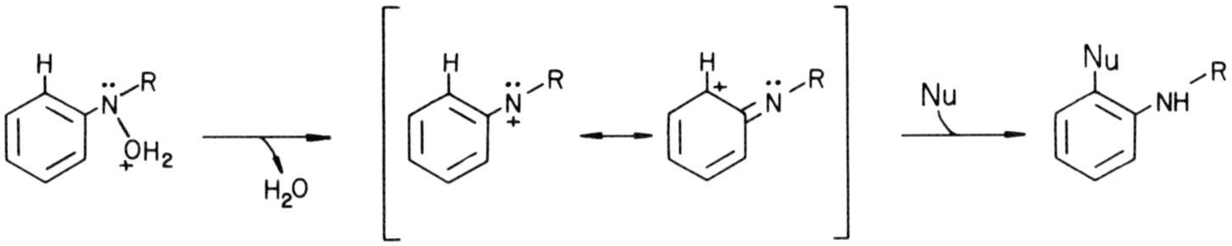

Fig. 1. The formation of the nitrenium ion from aromatic hydroxylamines as postulated by Booth and Boyland [1964]. Nu is a nucleophile such as OH^-, protein or nucleic acid.

E. Nitro-Compounds

The *C*-nitro-compounds are poorly soluble in water and have appreciably high boiling points [Smith, 1966]. If they possess an α-hydrogen, the protons are acidic and the compounds display a tautomeric equilibrium with nitronic acid derivatives. These compounds commonly decompose at moderate temperatures to give nitrous acid, HNO_2. *C*-Nitro-derivatives are relatively easily oxidized to give nitrous acid and ketones via radical intermediates. They are easily reduced in protic solvent and almost any reduction state can be detected including the nitro-anion radical. The nitro-anion radical can react with molecular oxygen to form superoxide anion radical and the parent nitro-compound. In addition, dimeric products of the reduction process, such as azoxy-, azo-, and hydrazine derivatives, are seen with aromatic nitro-compounds, but not aliphatic nitro-compounds.

F. Azo- and Diazene Derivatives

Due to the instability of the postulated *N*-hydroxyhydrazines formed by *m*-chloroperbenzoic acid oxidation of hydrazines or reduction of *N*-nitrosamines, the azo- and diazene derivatives most likely are the stable dehydration products of such oxidation reactions. The degree and position of substitution on the adjacent unsaturated nitrogens are determining factors in the stability of the compounds [Smith, 1966]. For example, 1,1-disubstituted diazenium ions are transiently stable and generally undergo one of two reactions [McBride and Kruse, 1957; Lemal et al., 1964]: dimerization to form tetrasubstituted tetrazenes or 2,3-sigmatropic shifts to form hydrazones. 1,2-Disubstituted azo-derivatives are more stable unless the compound possesses an α-hydrogen. In this case, they can tautomerize to form the hydrazone derivative; the equilibrium of the tautomerization reaction with benzyl-substituted azo-compounds lies in the direction of the hydrazone derivative [Simon and Moldenhauer, 1967]. These derivatives can exist as *cis/trans*-isomers; the *cis*-isomers are thought to be less stable. The *cis*-isomers are prone to undergo elimination of molecular nitrogen and form carbon-centered radicals at a faster rate than the *trans*-isomers [Smith, 1966]. 1,2-Disubstituted azo-derivatives can be reduced back to the hydrazine derivative (and hydrazyl radicals). Oxidation of these azo-compounds by peracids gives good yields of the azoxy derivatives.

Diazenes, monosubstituted azo-compounds, are thought to decompose rapidly to give molecular nitrogen and the parent hydrocarbon via a radical recombination reaction [Huang and Kosower, 1968; Kosower, 1971]. The alkyl- or aryl-radicals are reactive and can participate in typical radical reactions, such as addition, fragmentation, and rearrangement. Oxidation of diazenes, most likely, leads to formation of diazonium ions which can subsequently decompose to give carbonium ions [Smith, 1966].

G. Azoxy Derivatives

Like azo-compounds, these compounds exist as *cis/trans*-isomers, and the *trans*-isomers are thought to be more stable chemically [Smith, 1966]. The N-O bond can form with either nitrogen depending on the electron-withdrawing properties of the carbon substituents. Like *N*-oxides, they can rearrange in such a manner that the oxygen migrates from nitrogen to carbon. This reaction is commonly known as the Wallach rearrangement. In addition, the oxygen can be transferred to adjacent aliphatic carbons to form hydrazones or to other molecules in an intermolecular oxidation reaction [Smith, 1966]. In the second case, the azoxy-derivative is converted to its azo-parent.

II. Biological Reactivity of Oxidized Nitrogen Derivatives

A. Amine Oxides

The biological formation of these compounds is discussed in detail in chapters III and IV. Little information exists documenting any toxicity related directly to the *N*-oxide functional group. Several nitroquinoline *N*-oxides are carcinogenic agents, but considerable evidence indicates that it most likely is the nitro-group which, upon reduction to an arylhydroxylamine, is reactive (see below). Due to the extreme water-solubility of *N*-oxides, these compounds are rapidly excreted as urinary products [Lintzel, 1934]. In vivo and in vitro they are known to undergo at least two metabolic steps to yield either the parent amine upon reduction [Suguira et al., 1976] or the desalkyl amine and an aldehyde by a dealkylation reaction [Machinist et al., 1966]. This dealkylation reaction is similar to the Polonovski reaction [Smith, 1966]. The extensive further metabolism is most likely responsible for only partial recovery of *N*-oxides from animals injected with amine oxides [Dehner et al., 1968]. Recently, an *N*-oxide reductase has been described for *N,N*-dimethyl-*p*-aminoazobenzene; this purified protein was shown to be a cytosolic, NADPH-dependent *N*-oxide reductase [Johnson and Ziegler, 1986].

B. Hydroxylamines

The hydroxylamines are probably the best studied of the biological *N*-oxidation products (chapters III–V). Historically, it is well known that aromatic hydroxylamines, such as phenylhydroxylamine, cause methemoglobinemia in vivo [Heubner et al., 1923]. It was shown that phenylhydroxylamine is the N-oxidation product of aniline derivatives and reacts with oxyhemoglobin to form methemoglobin. A number of agents are known to react with oxyhemoglobin to form methemoglobin and an oxidized product [Eyer et al., 1975]. Many of these reactions apparently involved H_2O_2 production which may subsequently oxidize the ferrous iron of the heme protein. Aromatic hydroxylamines appeared to cause ferrihemoglobin formation concomitant with H_2O_2 production, but did not involve H_2O_2 or O_2 as intermediates in the reaction. The reaction with phenylhydroxylamine occurred at much higher rates of reaction than H_2O_2 itself, and exogenously added catalase or superoxide dismutase had no effect on the reaction. During oxyhemoglobin oxidation, the hydroxylamine appeared to be oxidized to nitrosobenzene.

Recently, Wallace et al. [1982] and Wallace and Caughey [1975] provided evidence for two mechanisms for oxyhemoglobin oxidation. Nonoxidizable nucleophiles, such as azide, thiocyanate, and fluoride ion, displaced superoxide anion from a protonated heme iron (Fe^{2+})-oxygen species to give ferric (Fe^{3+}) hemoglobin. However, oxidants, like phenylhydroxylamine, reacted by apparently donating an electron to bound dioxygen (reductive displacement) to give hydrogen peroxide and presumably nitroxyl radical (fig. 2). The nitroxyl radical can either react with another Fe^{2+}-O_2 molecule or with oxygen to form superoxide anion

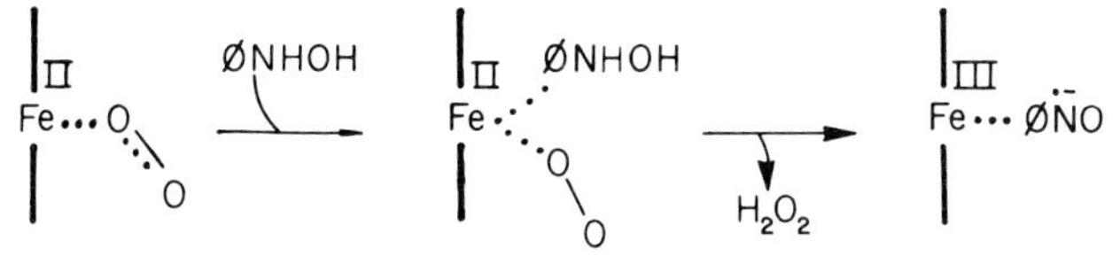

Fig. 2. The reductive displacement of oxygen from oxyhemoglobin as proposed by Wallace et al. [1982].

radical. The conclusion of Wallace and Caughey [1975] was that two mechanisms exist for hemoglobin auto-oxidation: proton-assisted nucleophilic displacement to form superoxide anion and reductive displacement to form hydrogen peroxide. In the red blood cell, the hydroxylamines can seriously impair oxygen transport to tissues by facilitating methemoglobin formation by the reductive displacement mechanism. Unless the red cell has adequate protection (i.e. adequate methemoglobin reductase activity) to maintain high steady-state concentrations of oxyhemoglobin, the aromatic hydroxylamines can cause serious problems in tissue oxygenation.

The aromatic hydroxylamines are nucleophilic and have been shown to serve as ligands to a number of cytochromes. One such example is the effect of the metabolites of alkylamines on the function of the microsomal cytochrome P-450 [Schenkman et al., 1972; Franklin, 1974; Mansuy et al., 1979]. Compounds, such as benzphetamine, are metabolized to *N*-hydroxybenzylamphetamine or *N*-hydroxyamphetamine which subsequently forms abortive complexes with the cytochrome, terminating the function of the monooxygenase. However, certain of the complexes formed can be decomposed by chemical means to regenerate the native form of the cytochrome. It is assumed that the action of alkylamines, such as SKF 525A and benzphetamine, on cytochrome function is due to the reactions of hydroxylamines and may account for certain drug-drug interactions seen with these compounds [Mansuy et al., 1979].

In addition, it is known that aromatic hydroxylamines interact with cellular nucleophiles, such as protein, thiols, and nucleic acids. Based on the early work of Booth and Boyland [1964], who linked the formation of *o*- and *p*-aminophenols to the reactivity of aromatic hydroxylamines (fig. 1), it was assumed that the -OH (or $-OH_2^+$) is a good leaving group which can allow the formation of a nitrenium cation. The positive charge can be delocalized to the aromatic ring. This electrophilic species can react with nucleophiles, such as OH^-, glutathione, protein thiols, and

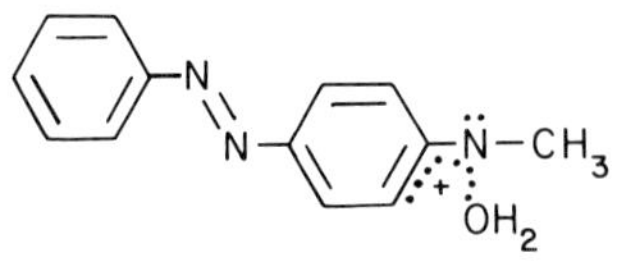

Fig. 3. The putative reactive species of aminoazo dyes as envisioned by Miller [1978] and Ketterer et al. [1982].

amino functions, to form covalent adducts with the aromatic nucleus of the arylamines.

The carcinogenic arylamines have been suggested to be converted to aromatic hydroxylamines which can form the putative nitrenium cation and react with cellular nucleophiles. Miller and Miller [1947] had shown in the late 1940s that aminoazo-dyes covalently bind to protein, lending their bright yellow color to the bound proteins. Subsequently, they have shown that these compounds, particularly the *N*-methylaminoazo-dyes, are *N*-oxidized to hydroxylamines. These *N*-hydroxylaminoazo-dyes then react with cellular nucleophiles, such as glutathione, protein, and DNA (fig. 3). These compounds are not acutely toxic in tissues such as liver. This is due most likely to the presence of detoxification pathways, such as sulfation, glucuronidation, and mercaptide formation, or the solvolysis of the reactive nitrenium ion by water to form excretable phenols [Ketterer et al., 1982]. However, sulfation or glucuronidation of the N-OH group can provide a better leaving group than -OH_2^+, but would lead to the formation of the reactive nitrenium ion [Irving, 1971; Kadlubar et al., 1976; Ketterer et al., 1982; Hammons et al., 1985].

C. N-*Hydroxyamides*

The reactivity of aryl *N*-hydroxyamides is similar to that described for the *N*-hydroxylamines. Several *N*-acetyl arylamines, such as *N*-acetylaminofluorene, are potent carcinogens [Cramer et al., 1960] and a classic example of the reactive intermediate formed is *N*-hydroxy-N-acetylaminofluorene (fig. 4). The formation of a nitrenium ion due to the presence of suitable leaving groups (-OH_2^+, -OSO_3H, or -O-glucuronide) has been shown to lead to either phenol formation or covalent bond formation between proteins, thiols, or nucleic acid and the aromatic nucleus of the hydroxamic acid [Irving et al., 1971; Kadlubar et al., 1976; Lai et al., 1985]. Some conflict exists in the literature related to the ability of the *N*-acetyl aromatic amides and aromatic amines to be interconverted, their capacity to be *N*-hydroxylated, their relative rates of acetylation/

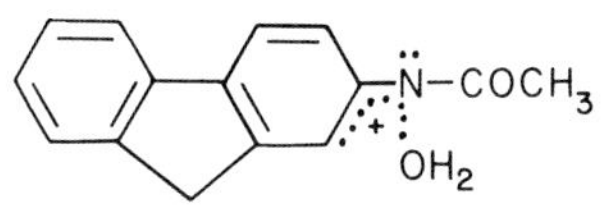

Fig. 4. The putative reactive species of *N*-hydroxyacetylamino-fluorene as envisioned by Miller [1978].

deacetylation, and their relative reactivity as electrophilic reagents [Miller, 1978].

The metabolic activation of other hydroxamic acids can also lead to toxic reactions. Prime examples are acetaminophen and phenacetin which are hepatotoxic under certain conditions. In a series of classic papers, Mitchell and co-workers described the role of metabolism of acetaminophen in its hepatotoxic activity [Mitchell et al., 1973a,b; Jollow et al., 1973; Potter et al., 1973]. It was demonstrated that upon metabolism, acetaminophen could arylate hepatic proteins in vivo and in vitro. A glutathione conjugate was noted and subsequent studies demonstrated that glutathione protected hepatocytes from toxicity until the glutathione levels were depleted. Based on other studies, it would appear that acetaminophen and other acetanilide derivatives can be *N*-hydroxylated by the cytochrome-P-450-dependent monooxygenase [Hinson et al., 1975]. The *N*-hydroxy-derivative of acetaminophen apparently is dehydrated to form an acetimidoquinone which can covalently bind to cellular nucleophiles, such as glutathione and protein. Some inconsistencies exist with this proposed mechanism, and an alternate pathway involving direct dehydrogenation of the *N*-acetyl-arylamine to a reactive iminoquinone has also been postulated. It is of interest that the *O*-ethyl derivative of acetaminophen, phenacetin, was activated apparently to a reactive intermediate by a pathway other than that involving acetaminophen [Hinson et al., 1977]; possibly by the formation of a 2,3-epoxide which rearranges subsequently to the iminoquinone.

D. N-*Nitrosamines*

While the rearrangement and reduction of nitrosamines would lead to formation of diazohydroxide intermediates which decompose to form carbonium ions, little evidence exists for such a pathway in biological systems. However, the work of Hecht et al. [1978] and Chen et al. [1978] has suggested that α-hydroxylation of nitrosamines accounted for the metabolic activation to alkyldiazohydroxides which, as expected above,

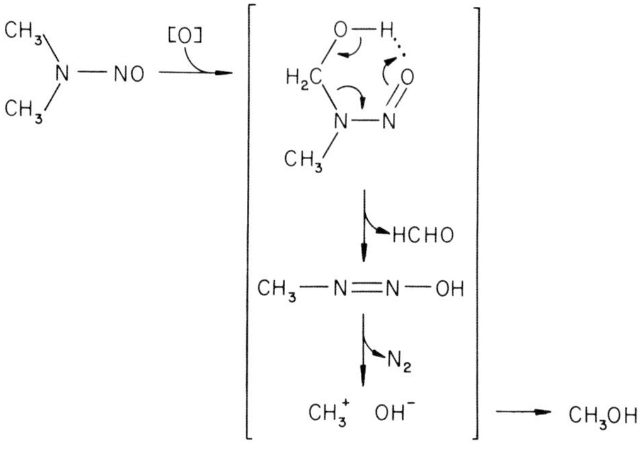

5

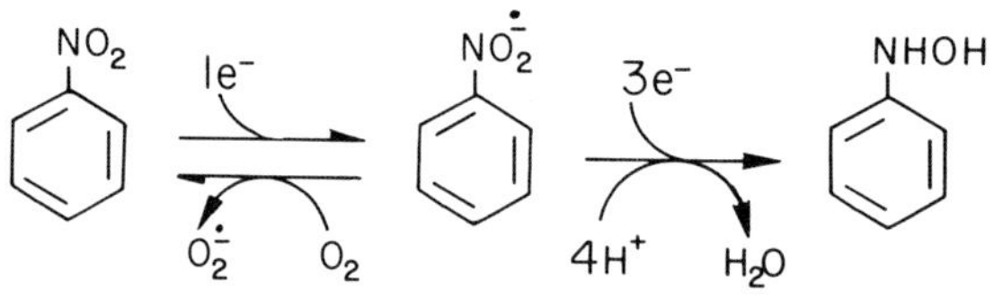

6

Fig. 5. The α-hydroxylation reaction envisioned in the metabolic activation of nitrosamines [Hecht et al., 1978].

Fig. 6. The reductive reactions of nitro compounds by biological systems [Boyd, 1980; Holtzman, 1982].

leads to formation of carbonium ions (fig. 5). Therefore, the nitrogen-oxygen bond is not directly activated by *N*-oxidation.

E. Nitro-Compounds

Two principal phenomenon leading to the deleterious effects of nitro-compounds have been proposed (fig. 6). Nitro-compounds have been shown to be reduced by flavin-dependent systems to hydroxylamines and amines. In the case of the carcinogen, 4-nitroquinoline *N*-oxide [Kato et al., 1970; Tada and Tada, 1976], the *N*-oxide was absolutely required as a structural component, but it was the nitro-group which was reductively activated by forming the hydroxylamine derivative. It is presumed that the 4-hydroxylaminonitroquinoline *N*-oxide reacts like the arylamines in forming a nitrenium ion.

The nitrofuran derivatives have been employed as bacteriostatic agents, but their usage has been reduced since many have been shown to be carcinogenic [Holtzman, 1982]. For example, 5-nitrofuran was used as a food additive in Japan prior to the realization that the agent was carcinogenic. Nitrofurantoin, one of the major commercial nitrofuran derivatives, is used as an antibacterial against urinary infection in man and animals [Stiko and Sasame, 1978]. It has also been shown to be a lung and livertoxin in certain species. Based on the known chemical reactions of aromatic nitro-compounds and furans, three possible routes of metabolic activation can be envisioned: epoxide formation, reduction to the arylhydroxylamine, and reduction to the air-sensitive nitro-anion radical. Boyd and co-workers demonstrated that covalent binding of nitrofurantoin to protein was increased in vitro under anaerobic conditions, but that covalent binding in vivo was not greater in liver tissue compared to lung [Boyd et al., 1980]. It was deduced that the formation of lung-specific epoxides of the furan ring, as seen for ipomeanol, was not a significant mechanism of nitrofurantoin toxicity. In certain tissues, reduction to the hydroxylamine may account for the covalent binding seen in vitro under anaerobic conditions. The formation of the hydroxylamine and, subsequently, the putative nitrenium ion, would be favored by anaerobiosis since the first reduction product, the nitro-anion radical, has been shown to be rapidly oxidized by molecular oxygen (fig. 6).

Under aerobic conditions, this radical species could not be observed, and reduction to hydroxylamine was prevented due to the air sensitivity of the nitro-anion radical [Mason and Holtzman, 1975]. The lung is a highly oxygenated organ, and the lung toxic properties of nitrofurantoin have been ascribed to an 'oxidative stress' similar to that seen with paraquat [Boyd, 1980]. The redox cycle involving the nitro-anion radical has been shown to generate appreciable amounts of hydrogen peroxide and superoxide radical in vitro. This mechanism is supported by the enhanced mortality in selenium-deficient chicks given nitrofurantoin orally [Peterson et al., 1980]. The observation implicated glutathione peroxidase in the protection against nitrofurans by maintaining low levels of intracellular hydrogen peroxide. However, it is of interest that vitamin E deficiency did not enhance the 'oxidative stress' of nitrofurantoin. In addition, the liver levels of reduced glutathione in chicks treated with nitrofurantoin were reduced relative to the controls. It is assumed that a number of flavin-dependent systems can reduce the nitro-compounds to the nitro-anion radical or hydroxylamine derivatives [Mason and Holtzman, 1975; Tada and Tada, 1976]. While no direct evidence for 'oxidative stress' in pulmonary toxicity exists, the data presented above require further examination of the possible toxicity of nitro-compounds by mechanisms involving redox cycling to form reactive species of oxygen.

F. Azo- and Diazene Derivatives

These derivatives represent the products of hydrazine oxidation or *N*-nitrosamine reduction and are most likely the first products which have appreciable chemical stability. The 1,2-disubstituted azo-compounds have been noted to be stable under physiological conditions and significant amounts of azo-methane and azo-procarbazine are accumulated in blood of animals treated with either 1,2-dimethylhydrazine or procarbazine, a 2-methylbenzylhydrazine. Three principal enzyme systems have been observed to *N*-oxidize the hydrazines to their azo-derivatives: microsomal cytochrome P-450 [Dunn et al., 1979; Prough et al., 1982], mitochondrial monoamine oxidase [Coomes and Prough, 1983; Erikson and Prough, 1986], and the microsomal FAD-containing monooxygenase [Prough et al., 1981]. Little evidence exists to suggest that these stable azo-products are toxic, but they most likely are further metabolized to form reactive free radicals or carbonium ions via *C*-hydroxylated intermediates.

While the monosubstituted diazenes may have measurable lifetimes under certain conditions, they are much less stable than the 1,2-disubstituted azo-derivatives. Several fates have been suggested for these intermediates. Further oxidation of alkyldiazenes may lead to alkyl diazonium ions which can yield carbonium ions [Cummings et al., 1982]. As seen in the section dealing with *N*-nitrosamines, the carbonium ions formed can react with cellular nucleophiles to yield covalent adducts or be solvolyzed to give alcohols (fig. 7). In addition, the diazenes are noted to decompose to form radicals which can also react with a number of cellular nucleophiles. Other products representative of radical cage recombination reactions can be detected, such as hydrocarbon products. An example of such recombination reaction is seen in figure 7. Erikson and Prough [1986] have demonstrated that diazene intermediates are the preferred metabolic products of monoalkylhydrazines when monoamine oxidase catalyzes their *N*-oxidation, but metabolism by cytochrome P-450 preferentially yields hydrazone derivatives as products. The difference in product ratio supports the contention that cytochrome P-450 tends to remove electrons from carbon centers rather than nitrogen centers [Frederick et al., 1982; Guengerich et al., 1982].

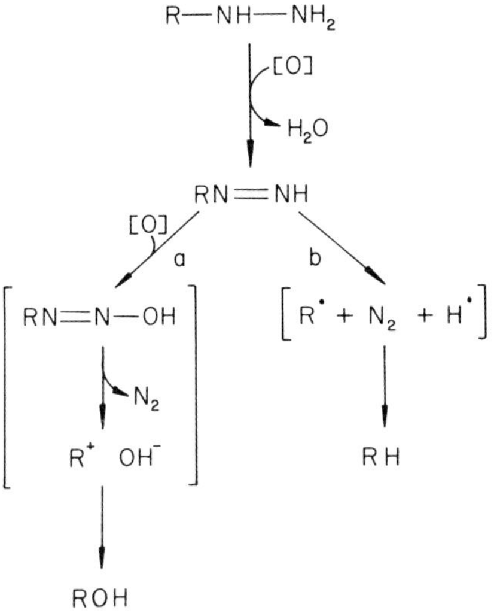

Fig. 7. The postulated mechanism of activation to alkyldiazenes by biological systems [Cummings et al., 1982; Moloney et al., 1985]. *a* Diazohydroxide path; *b* free radical path.

Recently, Moloney and Prough [1983] have reported that metabolism of procarbazine can lead to the formation of methane derived from the *N*-methyl group of the compound. Of the biological products of procarbazine metabolism, the azo-derivative appeared to be the major proximal substrate for methane production. Methylhydrazine and the hydrazone tautomer of azo-procarbazine were formed in insufficient quantities to account for the bulk of methane production [Moloney and Prough, 1983]. The azoxy-metabolites were not metabolized to methane either by reaction mixtures containing microsomal protein and NADPH or in isolated hepatocytes.

The methyl group of procarbazine and azo-procarbazine could also be metabolically activated in the presence of liver microsomal protein and NADPH to form a species capable of alkylating microsomal protein and DNA [Moloney et al., 1985]. The azoxy-procarbazine metabolites were not metabolically activated by liver microsomal fractions [Prough et al., 1982]. In order to relate the phenomena of methane formation and protein methylation, a series of induction and inhibitor studies were performed. The study demonstrated that both reactions were mediated by cytochrome P-450 and that the reactions behaved in similar ways to the presence of inhibitors of cytochrome P-450 and to animal pretreatment with either phenobarbital or 5,6-benzoflavone. It was noted that methane production could be stimulated by addition of thiol compounds such as glutathione, while protein binding decreased concomitantly. These results suggested that a reactive methyl free radical may be formed which could be converted to methane by thiols or could react with proteins to form covalent adducts. While the ramifications of this observation are not clear, the methyl radical may represent one reactive intermediate formed during the N-oxidation of hydrazine and azo-derivatives.

The best studied example of diazene reactions is that of phenyldiazene. It was observed that phenylhydrazine can cause hemolytic anemia by destroying hemoglobin [Lemberg and Legge, 1949]. Since it has been demonstrated that the reaction of phenylhydrazine and oxyhemoglobin yielded H_2O_2, it could be concluded that phenylhydrazine, like phenylhydroxylamine, can cause the reductive decomposition of oxyhemoglobin and can be oxidized to phenyldiazene. Itano and Mannen [1976] initially described this reaction as involving a form of phenylhydrazine which was covalently bound to the hemoglobin. Saito and Itano [1981] and Ortiz de Montellano and Kunze [1981] have subsequently demonstrated that the oxidation product of phenylhydrazine, most likely phenyldiazene or phenyl radical, can subsequently *C*- and *N*-phenylate the heme prosthetic group of hemoglobin. An iron-bound carbon radical is presumed to be an intermediate in the reaction [Mansuy et al., 1979]. This reaction accounted for the irreversible decomposition of oxyhemoglobin, instead of the potentially reversible formation of methemoglobin caused by agents such as azide or fluoride.

The oxidized products of hydrazines, in particular the unstable and air-sensitive diazenes, have been reported to bind to a number of heme compounds. Hines and Prough [1980] reported that 1,1-disubstituted hydrazine derivatives form metabolite complexes with cytochrome P-450 and terminate its function both in vivo and in vitro. This complex did not destroy the heme of the cytochrome, but upon depletion of the hydrazine substrate from solution or oxidation of the ferrous heme protein, the complex decomposed to give the native heme protein. Based on the chemistry of 1,1-disubstituted hydrazine and diazene derivatives [Lemal et al., 1964], it was assumed that the metabolite complex is composed of a 1,1-disubstituted diazenium ion bound as a ligand at the sixth coordination position of the heme of cytochrome P-450 (fig. 8).

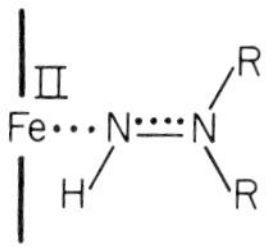

Fig. 8. The proposed structure of the inhibitory metabolite complex of cytochrome P-450 and an oxidative metabolite of 1,1-disubstituted hydrazines [Hines and Prough, 1980].

Jonen et al. [1982] subsequently studied the effect of phenylhydrazine on microsomal cytochrome P-450. It was noted that using microsomal suspensions in the presence of phenylhydrazine, oxygen, and NADPH, one initially could detect the loss of absorbance at 416 nm representing the loss of the low-spin form of the cytochrome. Subsequently, a unique spectral intermediate was observed with an absorbance maxima at 480 nm. The formation of absorbance at 480 nm was noted to occur concomitantly with the loss of the CO-reactive heme protein and a loss of heme measured as the pyridine hemochromagen. These results were interpreted to indicate that the cytochrome-P-450-dependent monooxygenase can catalyze the NADPH- and O_2-dependent oxidation of phenylhydrazine to phenyldiazene. The phenyldiazene can react subsequently with the heme of cytochrome P-450 to convert the CO-reactive heme protein to a protein containing degraded heme. Further experiments using ^{14}C-phenylhydrazine and liver microsomes demonstrated that the heme associated with cytochrome P-450 was alkylated, apparently in a manner similar to oxyhemoglobin [Moloney and Prough, unpublished results].

While the alkylation of cytochromes by diazenes may lead to inhibition of many heme-dependent reactions [Allison et al., 1973; Hidaka and Udenfriend, 1970], the toxicological effects of alteration of cytochrome P-450 are more subtle. The fact that hydrazines upon metabolism can alter monooxygenase function suggests that a number of deleterious drug-drug interactions may result [Hines and Prough, 1980; Moloney et al., 1984]. In fact, it was noted in the 1950s that iproniazid was an inhibitor of monooxygenase functions in vivo in a manner similar to SKF 525A [Fouts and Brodie, 1956].

G. Azoxy Derivatives

The azoxy compounds can be formed in biological systems in two ways: *N*-oxidation of stable azo-derivatives [Dunn et al., 1979; Wiebkin and Prough, 1980] and the condensation of hydroxylamines with their

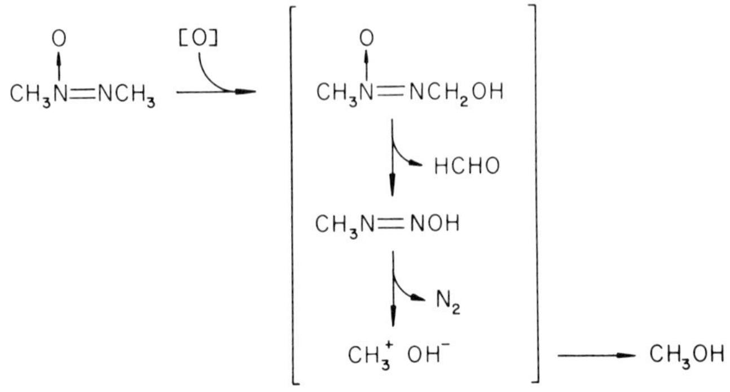

Fig. 9. The pathway of metabolic activation proposed for methylazoxymethanol [Druckrey, 1970; Fiala et al., 1978].

C-nitroso-oxidation products [Becker and Sternson, 1981; Fiala et al., 1981]. In these cases, the first is a direct enzymatic step and the second is a chemical coupling of two metabolites of the *N*-oxidation reactions of amines and the easily oxidized hydroxylamines. The azoxy compounds themselves are not highly reactive, but are thought to be subsequently activated by a *C*-hydroxylation reaction [Druckrey, 1970; Fiala et al., 1978; Weinkam and Shiba, 1978]. The hydroxyazoxy metabolite is thought to decompose to give an aldehyde and a diazohydroxide (fig. 9). Since it is well established that alkyldiazohydroxides form alkyl carbonium ions [Smith, 1966], it has been assumed for methylazoxymethanol that the hydroxyazoxy compound gives methyl carbonium ion, nitrogen, and formaldehyde [Druckrey, 1970; Fiala et al., 1978]. The reactive carbonium ion can then react with nucleophiles such as water to form methanol or protein and DNA to form methyl adducts. This situation is very similar to that seen for the metabolic activation of *N*-nitrosamines. Studies currently in progress in this laboratory indicate that a cytosolic enzyme may be involved in the final metabolic activation of the azoxy derivatives [Tweedie et al., unpublished results].

III. Covalent Binding and Toxicity

Since the initial observations of Boyland [1950], Magee and Barnes [1967], and the Millers [1966] that biological systems may form reactive chemical intermediates which can covalently bind to cellular nucleo-

Table I. Approaches to relate metabolism and covalent-binding to toxicity

1. Demonstrate that the compound covalently binds to cellular nucleophiles in a model system
2. Establish sufficient information on the routes of metabolism to insure that covalent binding occurs, not reincorporation of radiolabel from the compound into the normal synthetic pathways of the cell
3. Use of inhibitors and inducing agents which alter covalent binding and metabolism in order to relate these phenomena to the toxic response
4. Demonstrate the chemical nature of the adducts formed with cellular nucleophiles such as glutathione, proteins or nucleic acids
5. Demonstrate that cellular defense mechanisms, such as glutathione levels, are altered in parallel with the toxic response by various in vivo manipulations

philes, the question of how covalent binding to DNA or protein is related to the carcinogenicity and/or toxicity of various chemicals has not been answered clearly. Only after three decades has sufficient evidence been amassed for the polycyclic aromatic hydrocarbon, benzo(a)pyrene, to describe in molecular detail how the molecule must be oxygenated by cytochrome P-450 to form the *trans*-7,8-oxide, hydrolyzed by epoxide hydratase, to give the *trans*-7,8-dihydrodiol, and oxygenated again by cytochrome P-450 to yield the highly reactive *trans*-7,8-dihydrodiol-9,10-epoxy-benzo(a)pyrene [Thakker et al., 1985]. The stereochemistry of this process has been established and the ability of this electrophile to react with DNA has been evaluated both in vitro and in vivo. Yet, a number of unanswered questions still exist: which sites on DNA must be altered to lead to neoplasia, is the target organ the site of the three metabolic processes, and what happens after alteration of DNA?

Based on the work of Boyland, the Millers, Magee, and many others, Mitchell et al. [1975] attempted to develop strategies to elucidate whether the metabolic activation of chemicals could lead to a toxic event. Several of these approaches are included in table I. Although each of these approaches will provide information related to the reactive intermediates formed, Gillette [1980] has also pointed out that these approaches may not link the reactive intermediates to the toxic response in more than a casual fashion. However, such studies might provide considerable information related to the conditions which favor or disfavor the expression of toxicity by a given chemical.

In this review, we have provided examples from the literature indicating that a number of *N*-oxygenated and *N*-oxidized reactive intermediates can occur upon metabolism of nitrogenous compounds and yet others, such as N-oxides, are relatively nontoxic and are easily excreted. It seems obvious from the other chapters in this book and the studies of the many workers cited in this chapter that each nitrogenous compound and *N*-oxidized product possess a separate set of chemical properties. One example is the diversity of covalent adducts seen with various cellular nucleophiles (GSH, sulfates, glucuronides, proteins, and nucleic acids) and either *N*-acetylaminofluorine or monomethyl-aminoazobenzene [Miller, 1978]; adducts can be noted associated with the aromatic ring, nitrogen, and *N*-methyl groups. These differences in chemical reactivity explain in part why Gillette [1980] has argued that the approach of Mitchell et al. [1975] is limited until a large number of criteria have been met (table I). Therefore, considerable effort is required to elucidate the important covalent adduct(s) responsible for the carcinogenic and/or toxic phenomena of interest.

References

Allison, W.S.; Swain, L.C.; Tracy, S.M.; Benitez, L.V.: The reactivation of lactoperoxidase and the acyl phosphatase activity of oxidized glyceraldehyde-3-phosphate dehydrogenase by phenylhydrazine and phenyldiimide. Archs Biochem. Biophys. *155:* 400–404 (1973).

Bamberger, E.: Über das Phenylhydroxylamin. Berichte *27:* 1548–1561 (1894).

Bamberger, E.; Leyden, P.: Weitere Beiträge zur Kenntnis des Dimethylanilinoxyds. Berichte *34:* 12–16 (1901).

Banfield, F.H.; Kenyon, J.: Constitution of the condensation product of β-phenylhydroxylamine and acetone. J. chem. Soc. *1926:* 1612–1629.

Becker, A.R.; Sternson, L.A.: Oxidation of phenylhydroxylamine in aqueous solution: a model study of the carcinogenic effect of primary arylamines. Proc. natn. Acad. Sci. USA *78:* 2003–2007 (1981).

Bodendorf, K.; Blinder, B.: Properties of amine oxides. Arch. Pharmacol. *287:* 326–328 (1954).

Booth, J.; Boyland, E.: Metabolism of polycyclic compounds. 5. Formation of 1,2-dihydroxy-1,2-dihydronaphthalenes. Biochem. J. *91:* 362–369 (1964).

Boyd, M.R.: Biochemical mechanisms in pulmonary toxicity of furan derivatives; in Hodgson, Bend, Philpot, Reviews in biochemical toxicology, vol. 2, pp. 71–101 (Elsevier, New York 1980).

Boyland, E.: The biological significance of metabolism of polycyclic compounds. Biochem. Soc. Symp. *5:* 40–54 (1950).

Chen, C.B.; Hecht, S.S.; Hoffman, D.: Metabolic α-hydroxylation of the tobacco-specific carcinogen *N'*-nitrosonornicotine. Cancer Res. *38:* 3639–3645 (1978).

Coomes, M.L.; Prough, R.A.: The microsomal metabolism of 1,2-disubstituted hydrazines, procarbazine and 1,2-dimethylhydrazine. Drug Metab. Disposit. *11:* 550–555 (1983).

Cramer, J.W.; Miller, J.A.; Miller, E.C.: *N*-Hydroxylation: a new metabolic reaction observed in the rat with the carcinogen 2-acetylfluorene. J. biol. Chem. *235:* 885–888 (1960).

Cummings, S.W.; Guengerich, F.P.; Prough, R.A.: The characterization of *N*-isopropyl-*p*-hydroxymethyl-benzamide formed during the oxidative metabolism of azo-procarbazine. Drug Metab. Disposit. *10:* 459–464 (1982).

Cyerman-Craig, J.; Dwyer, F.P.; Glazer, A.N.; Horning, E.C.: Tertiary amine oxide rearrangements. I. Mechanism. J. Am. chem. Soc. *83:* 1871–1878 (1961).

Dehner, E.W.; Machinist, J.M.; Ziegler, D.M.: Effect of SKF 525-A on the excretion of amine oxides by the rat. Life Sci. *7:* 1135–1146 (1968).

Druckrey, H.: Production of colonic carcinoma by 1,2-dialkylhydrazines and azoalkanes; in Burdette, Carcinoma of the colon and antecedent epithelium, pp. 267–279 (Thomas, Springfield 1970).

Dunn, D.L.; Lubet, R.A.; Prough, R.A.: The oxidative metabolism of *N*-isopropyl-α-(2-methylhydrazino)-*p*-toluamide hydrochloride (procarbazine) by rat liver microsomes. Cancer Res. *39:* 4555–4563 (1979).

Dunstan, W.R.; Goulding, E.: The action of alkyl haloids on hydroxylamines. Formation of substituted hydroxylamines and oxamines. J. chem. Soc. *75:* 1004 (1899).

Erikson, J.M.; Prough, R.A.: Oxidative metabolism of some hydrazine derivatives by rat liver and lung tissues. J. Biochem. Toxicol. *1:* 41–52 (1986).

Eyer, P.; Hertle, H.; Kiese, M.; Klein, G.: Kinetics of ferrihemoglobin formation by some reducing agents, and the role of hydrogen peroxide. Molec. Pharmacol. *11:* 326–334 (1975).

Fiala, E.S.; Kohl, N.E.; Hecht, S.S.; Yang, J.J.; Shimada, T.: The formation of azoxy-2-phenylethane during the biological oxidation of phenylethylamine by rabbit liver microsomes. Carcinogenesis *2:* 165–173 (1981).

Fiala, E.S.; Kulakis, C.; Christiansen, G.; Weisburger, J.H.: Inhibition of the metabolism of the colon carcinogen, azomethane. Cancer Res. *38:* 4515–4521 (1978).

Fouts, J.R.; Brodie, B.B.: On the mechanism of drug potentiation by iproniazid (2-isopropyl-1-isonicotinyl hydrazine). J. Pharmac. exp. Ther. *116:* 480–485 (1956).

Franklin, M.R.: The influence of cytochrome P-450 induction on the metabolic formation of 455-nm complexes from amphetamines. Drug Metab. Disposit. *2:* 321–328 (1974).

Frederick, C.B.; Mays, C.B.; Ziegler, D.M.; Guengerich, F.P.; Kadlubar, F.F.: Cytochrome P-450 and flavin-containing monooxygenase-catalyzed formation of the carcinogen *N*-hydroxy-2-aminofluorene and its covalent binding to nuclear DNA. Cancer Res. *42:* 2671–2677 (1982).

Gillette, J.R.: An overview of the role of microsomal enzymes in the formation of toxic metabolites; in Coon, Conney, Estabrook, Gelboin, Gillette, O'Brien, Microsomes, drug metabolism, and chemical carcinogenesis, vol. II, pp. 777–789 (Academic Press, New York 1980).

Gilman, A.G., Jr.; Goodman, L.S.; Rall, T.W.; Murad, F.: in Goodman, Gillman, The pharmacological basis of therapeutics; 7th ed., pp. 1–1839 (Macmillan, New York 1985).

Guengerich, F.P.; Macdonald, T.L.; Burka, T.L.; Miller, R.E.; Liebler, P.C.; Zirvi, K.; Frederick, C.B.; Kadlubar, F.F.; Prough, R.A.: Diversified functions of cytochrome P-450; in Hietanen, Laitinen, Hanninen, Cytochrome P-450, biochemistry, biophysics, and environmental implications, pp. 27–34 (Elsevier, Biomedical Press, Amsterdam 1982).

Gutmann, A.: Über die Einwirkung von Stickoxyd- und Hydroxylamin-Verbindungen auf tert. Natriumarsenit. Berichte *55:* 3007–3012 (1922).

Hammons, G.J.; Guengerich, F.P.; Weis, C.C.; Beland, F.A.; Kadlubar, F.F.: Metabolic oxidation of carcinogenic arylamines by rat, dog, human hepatic microsomes and by purified flavin-containing and cytochrome P-450 monooxygenases. Cancer Res. *45:* 3578–3585 (1985).

Hecht, S.S.; Chen, C.B.; Hoffmann, D.: Evidence for metabolic α-hydroxylation of *N*-nitroso-pyrrolidine. Cancer Res. *38:* 215–218 (1978).

Heller, H.E.; Hughes, E.D.; Ingold, C.K.: A new view of arylhydroxylamine rearrangement. Nature *168:* 909–910 (1951).

Heubner, W.; Meier, R.; Rhode, H.: Studien über Methämoglobinbildung. V. Mitteilung: Phenylhydroxylamin. Arch. exp. Path. Pharmak. *100:* 149–161 (1923).

Hidaka, H.; Udenfriend, S.: Evidence of a hydrazine-reactive group as the active site of the non-heme portion of horseradish peroxidase. Archs Biochem. Biophys. *140:* 174–180 (1970).

Hines, R.N.; Prough, R.A.: The characterization of an inhibitory complex formed with cytochrome P-450 and a metabolite of 1,1-disubstituted hydrazines. J. Pharmac. exp. Ther. *214:* 80–86 (1980).

Hinson, J.A.; Mitchell, J.R.; Jollow, D.J.: Microsomal N-hydroxylation of *p*-chloroacetanilide. Molec. Pharmacol. *11:* 462–469 (1975).

Hinson, J.A.; Nelson, S.D.; Mitchell, J.R.: Studies on the microsomal formation of arylating metabolites of acetominophen and phenacetin. Molec. Pharmacol. *13:* 625–633 (1977).

Holtzman, J.L.: Role of reactive oxygen and metabolite binding in drug toxicity. Life Sci. *30:* 1–9 (1982).

Huang, P.-K. C.; Kosower, E.: Diazene. I. Decarboxylation of phenyldiazenecarboxylic acid. J. Am. chem. Soc. *90:* 2354–2362 (1968).

Irving, C.C.: Metabolic activation of *N*-hydroxy compounds by conjugation. Xenobiotica *1:* 387–398 (1971).

Itano, H.A.; Mannen, S.: Reactions of phenyldiazene and ring-substituted phenyldiazenes with ferrihemoglobin. Biochim. biophys. Acta *421:* 87–96 (1976).

Johnson, P.R.L.; Ziegler, D.M.: Properties of a *N,N*-dimethyl-*p*-aminoazobenzene oxide reductase purified from rat liver cytosol. J. Biochem. Toxicol. *1:* 15–27 (1986).

Jollow, D.J.; Mitchell, J.R.; Potter, W.Z.; Davis, D.C.; Gillette, J.R.; Brodie, B.B.: Acetaminophen-induced hepatic necrosis. II. Role of covalent binding in vivo. J. Pharmac. exp. Ther. *187:* 195–202 (1973).

Jonen, H.G.; Werringloer, J.; Prough, R.A.; Estabrook, R.W.: The reaction of phenylhydrazine with microsomal cytochrome P-450: catalysis of heme modification. J. biol. Chem. *257:* 4404–4411 (1982).

Kadlubar, F.F.; Miller, J.A.; Miller, E.C.: Hepatic metabolism of *N*-hydroxy-*N*-methyl-4-aminobenzene and other *N*-hydroxy arylamines to reactive sulfuric acid esters. Cancer Res. *36:* 2350–2359 (1976).

Kato, R.; Iwasaki, K.; Noguchi, H.: Stimulatory effect of FMN and methyl viologen on cytochrome P-450-dependent reduction of tertiary amino *N*-oxide. Biochem. biophys. Res. Commun. *72:* 267–274 (1976).

Kato, R.; Takahashi, M.; Oshina, T.: Characteristics of nitro reduction of the carcinogenic agent 4-nitroquinoline-*N*-oxide. Biochem. Pharmac. *19:* 45–55 (1970).

Ketterer, B.; Srai, S.K.S.; Wayforth, B.; Tullis, D.L.; Evans, F.E.; Kadlubar, F.F.: Formation of *N*-(gluthathione-*S*-methylene)-4-aminoazobenzene following metabolic oxidation of the *N*-methyl group of the carcinogen, *N*-methyl-4-aminoazobenzene. Chem.-Biol. Interact. *38:* 287–302 (1982).

Kosower, E.M.: Monosubstituted diazenes (diimides). Surprising intermediates. Acct. Chem. Res. *4:* 193–198 (1971).

Lai, C.C.; Miller, J.A.; Miller, E.C.; Liem, A.: N-Sulfooxy-2-amino-fluorene is the major ultimate electrophilic and carcinogenic metabolite of *N*-hydroxy-2-acetylaminofluorene in the livers of infant male C57BL/6J × C3H/HEJ F1 (B6C3F1) mice. Carcinogenesis *6:* 1037–1045 (1985).

Lemal, D.M.; Menger, F.; Coats E.: The diazene-hydrazone rearrangement. J. Am. chem. Soc. *86:* 2395–2401 (1964).

Lemberg, R.; Legge, J.W.; in Hematin compounds and bile pigments, pp. 1–524 (Intersciences, New York 1949).

Lintzel, W.: Untersuchungen über Trimethylammoniumbase. III. Trimethylammoniumbasen im menschlichen Harn. Biochem. Z. *273:* 243–261 (1934).

Machinist, J.M.; Orme-Johnson, W.H.; Zeigler, D.M.: Microsomal oxidases. II. Properties of a pork liver microsomal *N*-oxide dealkylase. Biochemistry *5:* 2939–2943 (1966).

Magee, P.N.; Barnes, J.A.: Carcinogenic nitroso compounds. Adv. Cancer Res. *10:* 163–246 (1967).

Mansuy, D.; Battioni, J.P.; Chottard, J.C.; Ullrich, V.: Preparation of a porphyrin-iron-carbene model for the cytochrome p-450 complexes obtained upon metabolic oxidation of the insecticide synergists of the 1,3-benzodioxole series. J. Am. chem. Soc. *101:* 3971–3973 (1979).

Mason, R.P.; Holtzman, J.L.: The role of catalytic superoxide formation in the O_2 inhibition of nitroreductase. Biochem. biophys. Res. Commun. *67:* 1267–1274 (1975).

McBride, W.R.; Kruse, H.W.: Alkylhydrazines. I. Formation of a new diazo-like species by the oxidation of 1,1-dialkylhydrazines in solution. J. Am. chem. Soc. *79:* 572–576 (1957).

Miller, E.C.: Some current perspectives on chemical carcinogenesis in humans and experimental animals: presidential address. Cancer Res. *38:* 1479–1496 (1978).

Miller, E.C.; Miller, J.A.: The presence and significance of bound aminoazo dyes in the livers of rats fed *p*-dimethylaminoazobenzene. Cancer Res. *7:* 468–480 (1947).

Miller, E.C.; Miller, J.A.: Mechanisms of chemical carcinogenesis: nature of proximate carcinogens and interactions with macromolecules. Pharmacol. Rev. *18:* 805–838 (1966).

Mitchell, C.H.; Ziegler, D.M.: A quantitative micro method for the estimation of amine oxides. Analyt. Biochem. *28:* 261–268 (1969).

Mitchell, J.R.; Jollow, D.J.; Potter, W.Z.; Davis, D.C.; Gillette, J.R.; Brodie, B.B.: Acetaminophen-induced hepatic necrosis. I. Role of drug metabolism. J. Pharmac. exp. Ther. *187:* 185–194 (1973a).

Mitchell, J.R.; Jollow, D.J.; Potter, W.Z.; Gillette, J.R.; Brodie, B.B.: Acetaminophen-induced hepatic necrosis. IV. Protective role of glutathione. J. Pharmac. exp. Ther. *187:* 211–217 (1973b).

Mitchell, J.R.; Potter, W.Z.; Hinson, J.A.; Snodgrass, W.R.; Timbrell, J.A.; Gillette, J.R.: Toxic drug reactions; in Hb. exp. Pharmakol., pp. 383–419 (Springer, New York 1975).

Moloney, S.J.; Prough, R.A.; Studies on the pathway of methane formation from procarbazine, a 2-methylbenzylhydrazine derivative, by rat liver microsomes. Archs. Biochem. Biophys. *221:* 577–584 (1983).

Moloney, S.J.; Snider, B.J.; Prough, R.A.: The interactions of hydrazine derivatives with rat-hepatic cytochrome p-450. Xenobiotica *14:* 803–814 (1984).

Moloney, S.J.; Wiebkin, P.; Cummings, S.W.; Prough, R.A.: Metabolic activation of the terminal *N*-methyl group of *N*-isopropyl-α-(2-methylhydrazino)-*p*-toluamide hydrochloride (procarbazine). Carcinogenesis *6:* 397–401 (1985).

Ortiz de Montellano, P.R.; Kunze, K.L.: Formation of *N*-phenylheme in the hemolytic reaction of phenylhydrazine with hemoglobin. J. Am. chem. Soc. *103:* 6534–6536 (1981).

Peterson, F.J.; Mason, R.P.; Holtzman, J.L.: The effect of selenium and vitamin E deficiency on the toxicity of nitrofurantoin in the chick; in Coon, Conney, Estabrook, Gelboin, Gillette, O'Brien, Microsomes, drug oxidation and chemical carcinogenesis, vol. II, pp. 873–876 (Academic Press, New York 1980).

Potter, W.Z.; Davis, D.C.; Mitchell, J.R.; Jollow, D.J.; Gillette, J.R.; Brodie, B.B.: Acetaminophen-induced hepatic necrosis. III. Cytochrome P-450-mediated covalent binding in vitro. J. Pharmac. exp. Ther. *187:* 203–210 (1973).

Prough, R.A.; Coomes, M.W.; Cummings, S.W.; Wiebkin, P.: Metabolism of procarbazine; in Snyder, Parke, Kocsis, Jollow, Gibson, Witmer, Biological reactive intermediates. 2. Chemical mechanisms and biological effects, pp. 1067–1076 (Plenum Press, New York 1982).

Prough, R.A.; Freeman, P.C.; Hines, R.N.: The oxidation of hydrazine derivatives catalyzed by the purified liver microsomal FAD-containing monooxygenase. J. biol. Chem. *256:* 4178–4184 (1981).

Saito, S.; Itano, H.A.: β-meso-phenylbiliverdin IXa and *N*-phenylprotoporphrin. IX. Products of the reaction of phenylhydrazone with oxyhemoproteins. Proc. natn. Acad. Sci. USA *78:* 5508–5512 (1981).

Schenkman, J.B.; Wilson, B.J.; Cinti, D.L.: Diethylaminoethyl 2,2-diphenylvalerate HCl (SKF-525)- in vivo and in vitro effects of metabolism by rat liver microsomes-formation of an oxygenated complex. Biochem. Pharmac. *21:* 2373–2383 (1972).

Simon, H.; Moldenhauer, W.: Untersuchungen zur Hydrazon-azo- und en-Hydrazin Tautomerie sowie entsprechender Tautomerien bei Semicarbazonen durch Wasserstoffaustausch. Chem. Ber. *100:* 1949–1960 (1967).

Smith, P.A.S.: Hydrazine derivatives; in The chemistry of open-chain organic nitrogen compounds, vol. 2, chap. 9, pp. 119–209 (W.A. Benjamen, New York 1966).

Stiko, A.; Sasame, H.A.: Metabolic activation of nitrofurantoin-possible implications for carcinogenesis. Biochem. Pharmac. *28:* 601–606 (1978).

Suguira, M.; Iwasaki, K.; Kato, R.: Reduction of tertiary amine *N*-oxides by liver microsomal cytochrome P-450. Molec. Pharmacol. *12:* 322–334 (1976).

Sulkowski, T.S.; Childress, S.J.: A new method for preparing 5-aryl-2,3-dihydro-1H-1,4-benzodiazepines. J. org. Chem. *28:* 2150–2156 (1963).

Tada, M.; Tada, M.: Metabolic activation of 4-nitroquinoline *N*-oxide and its binding to nucleic acid; in Magee, Takayama, Sugimura, Matsushima, Fundamentals of cancer prevention, pp. 217–227 (University Park Press, Baltimore 1976).

Thakker, D.R.; Yagi, H.; Levin, W.; Wood, A.W.; Conney, A.H.; Jerina, D.M.: Polycyclic aromatic hydrocarbons: metabolic activation to ultimate carcinogens; in Anders, Bioactivation of foreign compounds, pp. 178–242 (Academic Press, Orlando 1985).

Wallace, W.J.; Caughey, W.S.: Mechanism for the autoxidation of hemoglobin by phenols, nitrite, and 'oxidant' drugs. Peroxide formation by one electron donation to bound dioxygen. Biochem. biophys. Res. Commun. *62:* 561–567 (1975).

Wallace, W.J.; Houtchens, R.A.; Maxwell, J.C.; Caughey, W.S.: Mechanism of autoxidation for hemoglobins and myoglobins. Promotion of superoxide production by protons and anions. J. biol. Chem. *257:* 4966–4977 (1982).

Weinkam, R.J.; Shiba, D.A.: Metabolic activation of procarbazine. Life Sci. *22:* 937–946 (1978).

Wiebkin, P.; Prough, R.A.; Oxidative metabolism of *N*-isopropyl-α-(2-methylazo)-*p*-toluamide (azo-procarbazine) by rodent liver microsomes. Cancer Res. *40:* 3524–3529 (1980).

Williams, R.T.: in Detoxification mechanisms; 2nd ed., pp. 1–796 (Chapman & Hall, London 1959).

Ziegler, D.M.: Microsomal flavin-containing monooxygenase: oxygenation of nucleophilic nitrogen and sulfur compounds; in Jakoby, Enzymatic basis for detoxification, pp. 201–227 (Academic Press, New York 1980).

R.A. Prough, PhD, Department of Biochemistry, The University of Louisville School of Medicine, Louisville, KY 40292 (USA)

Prog. bas. clin. Pharmac., vol. 1, pp. 184–212 (Karger, Basel 1988)

VII. Metabolic Disposition of Nitrogen Functionalities

Arthur K. Cho

Department of Pharmacology, UCLA School of Medicine, Center for the Health Sciences, Los Angeles, Calif., USA

Introduction

This chapter considers the metabolic disposition of functional groups containing different oxidation states of nitrogen (table I). It will include the reactions described in earlier chapters but will also include reductive and synthetic metabolic reactions. The metabolic pathways available to any compound are determined by its functional groups so that this perspective of nitrogen metabolism should be useful in assessing the possible metabolites of new organic compounds. The metabolic transformation of the different nitrogen functionalities involve oxidation, reduction, hydrolysis and conjugation reactions, and each function is considered in terms of these types of reactions.

Amines

The amino-state is the most common oxidation state of nitrogen in compounds of pharmacological interest, and its usual metabolic transformation is by oxidation of the adjacent carbon atom to form the carbinolamine. Hydrolysis of this unstable functionality results in *N*-dealkylation, generating a nitrogen that is much more hydrophylic (reaction 1).

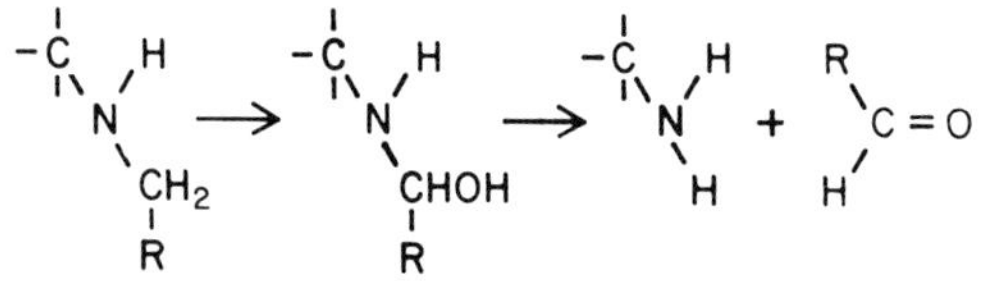

Table I. Nitrogen functional groups

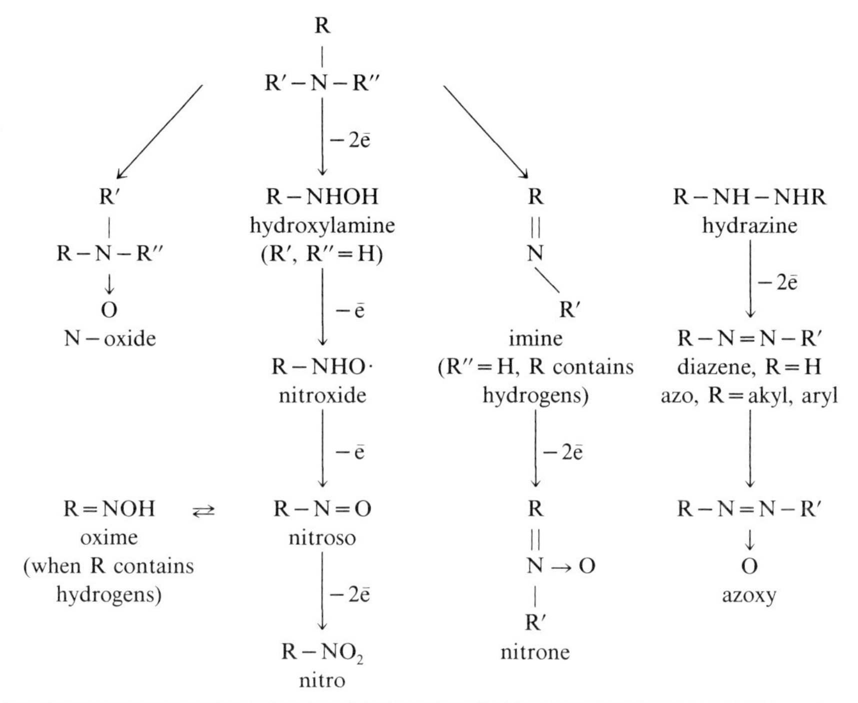

This carbon hydroxylation reaction has been extensively documented and reviewed elsewhere [37, 56, 64], but recent evidence described in chapter II indicates that the reaction can proceed through a one electron oxidation of the nitrogen. The nitrogen function can also retain oxygen-generating reaction products that depend on the substitution of the amine group. Thus, primary and secondary amines are converted to hydroxylamines, while tertiary amines are converted to *N*-oxides.

Oxidation

There are at least two enzyme systems capable of effecting two-electron oxidation of primary and secondary amines to hydroxylamines, the cytochrome P-450 system (chapter II) and the flavin monooxygenase (chapter IV). Amine *N*-hydroxylation was first described for the carcinogenic primary aromatic amines and their amides [16]. The responsible enzyme was proposed to be the cytochrome P-450 by Thorgeirsson et al.

[94] who concluded that the cytochrome P-450 system *N*-hydroxylates only aromatic amines and that arylalkylamines are *N*-hydroxylated by the flavin monooxygenase. Unfortunately, such a simple distinction was not the case, and in the following year Cho et al. [12] described the *N*-hydroxylation of phentermine (2-methyl-1-phenyl-2-propylamine), an arylalkylamine, by microsomes under conditions indicative of cytochrome P-450 involvement. Subsequently, the *N*-hydroxylation of primary aromatic amines [59] and arylalkylamines [21] by reconstituted cytochrome P-450 systems was reported. Secondary amines, such as *N*-ethylaniline, are also *N*-hydroxylated [96] and, although they appear to be more susceptible to the flavin monooxygenase, they are also oxidized by the P-450 system. Robertson et al. [82, 83] have presented data suggesting that one of the phenobarbital inducible cytochromes P-450, form 5, may be responsible for the *N*-hydroxylation of mutagenic aromatic amines such as 2-aminoanthracene and aminofluorene. However, their evidence for *N*-hydroxylation was based on a mutagenic response and not identification of a hydroxylamine. Frederick et al. [25] have studied the metabolism of 2-aminofluorene by microsomes from rat, dog, man and pig and by purified preparations of cytochrome P-450 and the flavin monooxygenase. These investigators assayed the products of *N*-oxidation and identified the hydroxylamine, the nitroso-, nitro- and azoxy derivatives.

N-hydroxylation was shown to be catalyzed by both cytochrome P-450 and the flavin oxygenase with the contributions of each varying with the species. Experiments with purified enzymes indicated that the *N*-hydroxylation was catalyzed by a native cytochrome P-450, not one induced by phenobarbital. In a related study Goldstein et al. [32] examined acetylaminofluorene metabolism by microsomes and reconstituted cytochrome P-450 systems and noted that 3-methylcholanthrene-inducible forms contributed only slightly to *N*-hydroxylation by rat liver microsomes. Tertiary aliphatic amines are converted to their *N*-oxides by the two oxidases as well [33]. Since the relative levels of these two *N*-oxidizing systems differ with species (chapter IV) and with the pharmacological history of an individual animal, their relative importance in the disposition of a particular compound cannot be readily predicted.

In other studies with primary aromatic amines, Kedderis and Rickert [49] demonstrated that 2-amino-6-nitrobenzyl alcohol, a metabolite of 2,6-dinitrotoluene, was hydroxylated at the amino nitrogen and the

5-position on the aromatic ring. The presence of the nitro-group appeared to increase microsomal oxidation of the benzyl alcohol, as 2-aminobenzyl alcohol was oxidized at about one fifth the rate. Procainamide, an amide of 4-aminobenzoic acid, is also *N*-hydroxylated at the 4-amino-position by a cytochrome-P-450-dependent pathway [97]. The *N*-hydroxy metabolite appears to be further metabolized rapidly, as very little was isolated in experiments with the perfused liver. Aromatic hydroxylamines are very unstable in the presence of oxygen so the disappearance could be the result of chemical as well as biochemical reactions.

Amino nitrogen can also undergo a single electron oxidation to a free radical. Aromatic amines such as paracetamol [66, 70] and *p*-phenetidine [66] have been shown to generate these reactive species.

In another odd electron oxidation process, cocaine has been shown to undergo *N*-oxidation at the tertiary nitrogen to form the *N*-oxide and the nitroxide, which is thought to be involved in the hepatic necrosis associated with this drug (for review, see ref. [53]).

Conjugative Reactions

Amines are conjugated by acylation or glycosidation. Acylation occurs with either formyl or acetyl groups but acetylation is more common. Formylation occurs with aromatic primary amines and appears to utilize kynurenine formamidase [85]. *N*-Acetylation is mediated by an acetyltransferase and acetyl coenzyme A [100]. The amine functionalities that are acetylated appear to be either aromatic amines or hydrazine derivatives. One of the reasons for the preferential acylation of these functional groups may be their relatively weak base strength. Aromatic amines, arylhydrazines and acylhydrazines are weak ($pK_a < 5$) bases so that at physiological pH they are essentially neutral compounds, whereas alkylamines with pK_a values of nine or greater are almost completely ionized. If transacetylation requires a neutral amino function, weak bases would be much better substrates.

Another transferase that catalyzes intramolecular N-O-transacetylation contributes to the hepatotoxicity of acylarylamines [52]. By translocating an acyl group from nitrogen to oxygen, the enzyme forms *N*-acetoxyarylamines, some of which are considered to be primary carcinogens (see chapter V). The amino group is also conjugated with sugars through a glycosidic bond (scheme 1). The nitrogen-carbon glycosidic

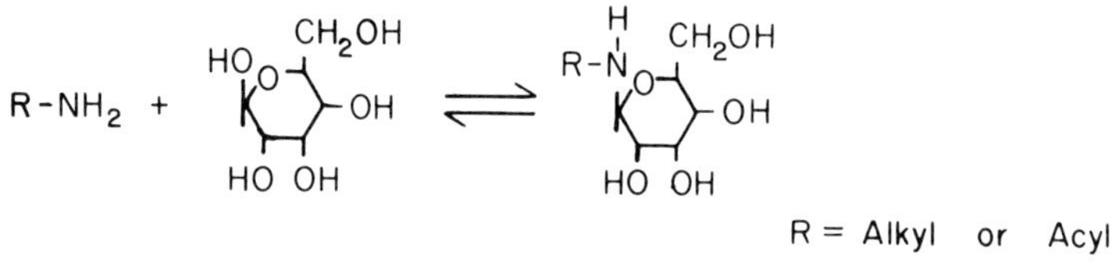

bond is unstable when the nitrogen is part of an amine but relatively stable when an amide is involved.

The instability of glycoside conjugates of amines makes their analysis difficult so that the importance of the pathway for amine disposition cannot be readily assessed. *N*-Glucuronide conjugates of amines are likewise unstable but are also readily formed so that the amino-glycosidic bond can form during the extraction of urine thereby generating a possible artifact. The amine glucuronide metabolite of cyproheptadine *(1)* is unusual in that it is a stable quaternary ammonium conjugate which is a major urinary excretion product [76].

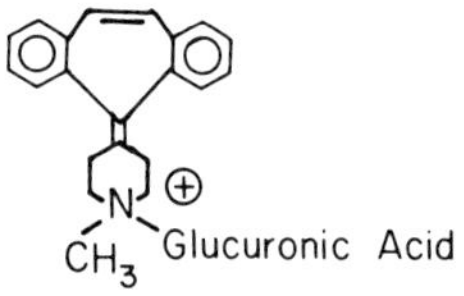

1

However, the highly polar nature of quaternary glucuronides may have caused them to be overlooked in other studies. Amide glucuronides have been reported as urinary metabolites of barbiturates [91] and of a triazole [20], but there appears to be little information on the biochemistry of these reactions.

Hydroxylamines and N*-Oxides*

These functionalities are two-electron oxidation products of amines and, depending on the structure, can undergo reductive, oxidative, and conjugative reactions. There are few drugs with hydroxylamine functions, but since hydroxylamines and *N*-oxides are common amine metabolites, their biotransformation is important in understanding amine metabolism.

Reduction

The reduction of aliphatic hydroxylamines and *N*-oxides is extensively documented and, like *N*-oxidation, appears to be due to several different enzymes. Kadlubar and co-workers [44, 45] have described an NADH-dependent, oxygen-insensitive hydroxylamine reductase in pig liver microsomes that reduces *N*-methyl-*N*-benzylhydroxylamine and *N*-methyl-1-octylhydroxylamine. In another study, Sum and Cho [90] demonstrated the reduction of *N*-hydroxyphentermine to phentermine by rat liver microsomes under anaerobic conditions. This oxygen-sensitive reaction is NADPH-dependent, inhibited by carbon monoxide and phenobarbital inducible, and thus appears to be cytochrome P-450 mediated. Since this latter reaction is oxygen-sensitive, its occurrence in the physiological state may be questioned. However, when administered to rats, *N*-hydroxyphentermine is rapidly reduced in vivo [11] so that some reductive pathway is available. This in vivo reduction may be due to a completely different enzyme system as yet characterized. On the other hand, free and conjugated hydroxylamines [3] are excreted as metabolites of phentermine so that reduction is not complete. More information is clearly needed to determine factors controlling the oxidation and reduction of hydroxylamines.

Tertiary amine *N*-oxides are also reduced by enzymes in the endoplasmic reticulum. Studies with microsomes and with reconstituted cytochrome P-450 systems [41, 89] indicate that cytochrome P-450 is partly responsible for *N*-oxide reduction. The *N*-oxides used in these

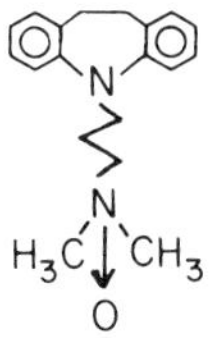

2

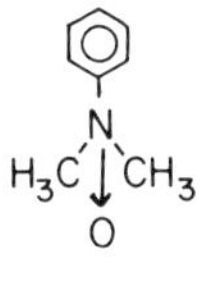

3

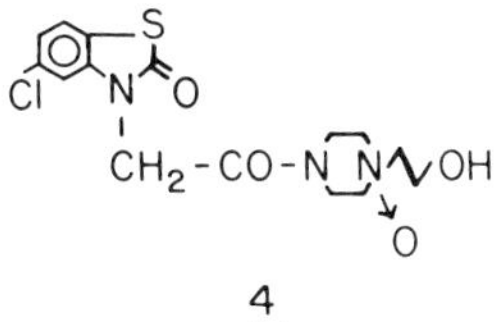

4

studies were imipramine *N*-oxide *(2)*, dimethylaniline *N*-oxide *(3)*, and tiaramide *N*-oxide *(4)*. Imipramine is a dimethylalkylamine, tiaramide a piperazine derivative, and dimethylaniline an aromatic amine, so that the reduction appears to be relatively nonspecific with respect to substrate structure. Microsomal reduction of nicotine *N*-oxide also occurs by an enzyme system that is sensitive to oxygen and carbon monoxide.

The reduction of aromatic hydroxylamines is not well documented, but indirect studies on aromatic nitro-compound reduction indicated that the intermediate hydroxylamines are formed and reduced to amines [48]. Kadlubar et al. [45] examined phenylhydroxylamine and 2-naphthylhydroxylamine as substrates for microsomal NADH-dependent reductase but measured substrate disappearance and assumed product reduction. When phenylhydroxylamine was used as substrate, both it and NADH were consumed, indicating a transformation had occurred while naphthylhydroxylamine did not oxidize NADH. In studies with 2-aminofluorene and 2-acetamidofluorene, Razzouk et al. [80] found that *N*-hydroxy metabolites *(5)* were rapidly metabolized further by guinea pig liver microsomes to 7-hydroxyacetaminofluorene *(6)*. The formation of this metabolite either involves a rearrangement of the hydroxylamine or its reduction and hydroxylation (scheme 2).

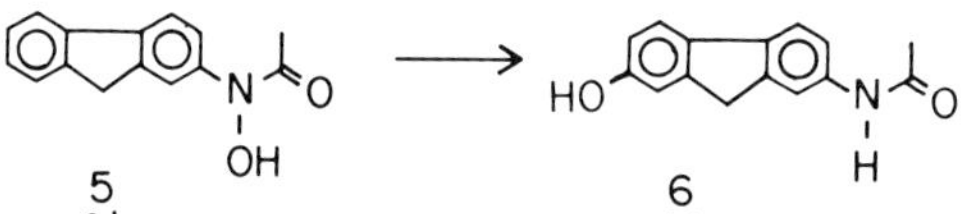

5 6

Oxidation

Oxidation to Nitroxides. A single electron oxidation of hydroxylamine yields the nitroxide radical which is usually unstable and rapidly disproportionates. However, stable nitroxides have been noted in microsomal incubates of aromatic hydroxylamines, such as *N*-hydroxyacetylaminofluorene, 2-aminoanthracene, and 4-amino-azobenzene among others [88]. Rauckman et al. [79] have used a highly hindered hydroxylamine, 2,2,6,6,-tetramethylpyridinoxyl, and demonstrated that stable radicals are generated by liver microsomes. This reaction is thought to be a superoxide (O_2^-) mediated oxidation with O_2^- generated by the flavin enzyme, NADPH amine oxidase.

Oxidation to Oximes. Aliphatic hydroxylamines with at least one α-hydrogen can be oxidized to oximes (chapter II). Phenylacetone oxime *(9)* was one of the first oximes to be reported and was identified in microsomal incubates of amphetamine *(7)* [38, 73, 103] and of *N*-hydroxyamphetamine *(8)* [103]. However, the oxime was not universally accepted as a metabolite of amphetamine. Beckett and Al-Sarraj [2]

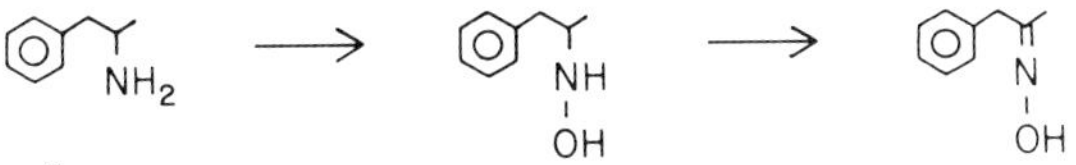

7 8 9

reported the metabolic conversion of amphetamine to *N*-hydroxyamphetamine, but not the corresponding oxime (reaction 2).

Since the oxime can form by chemical oxidation of the hydroxylamine, they proposed that it was a 'metabonate' or chemical decomposition product of the hydroxylamine metabolite. The discrepancy between the work of Beckett and others could be due to difference in the time of incubation. The hydroxylamine is an initial metabolite which can then be oxidized to the oxime so that at short incubation times only small amounts of oxime are formed. Beckett and Al-Sarraj [2] do not report their incubation time, and it could be much shorter than the ones used in the other studies which were all 60 min. or longer. A non-cytochrome P-450, NADPH/O_2-dependent enzyme system capable of converting *N*-hydroxyamphetamine to phenylacetone oxime has been described [62] indicating that the oxime can be formed enzymatically. The presence of oximes in metabolic incubates of other amines has been reported [13], but since the assumption in the study was that oximes were artifacts of the rigorous workup procedure, controls were lacking that would demonstrate enzymatic formation.

In a study of the metabolism of a series of α-substituted phenylethylamines by rabbit liver 9,000 *g* supernatant, Jonsson et al. [43] found oxime levels to be highest with the α-methyl compound (amphetamine) and substantially lower with α-ethyl-, α-isopropyl-, and α-tertiary butyl-β-phenylethylamine as substrates. All four compounds were metabolized to hydroxylamines, and except for amphetamine, levels of *N*-hydroxy metabolite increased with the size of alkyl group. A similar relationship has been observed for metabolic intermediate complex formation (chapter III).

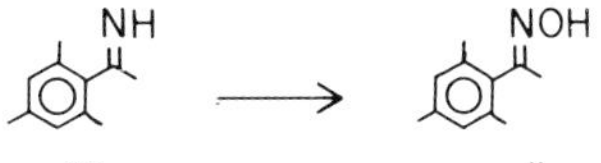

10 11

Another pathway for oxime formation has been described by Parli et al. [74], who incubated the highly hindered 2,4,6-trimethyl-acetophenone imine *(10)* with rat and rabbit liver microsomes and identified the oxime *(11)* as the major metabolite.

This reaction was carbon monoxide-, SKF-525A- and DPEA-sensitive so that cytochrome P-450 involvement seems likely. The reaction is unusual because imines are unstable and readily undergo hydrolysis. The substantial steric hindrance about the imino carbon presumably stabilizes this particular compound. Another example of a sterically hindered oxime *(13)* has been reported as a metabolite of dezocine *(12)*, a bridged aminotetralin derivative [87]. This oxime may also be a metabolite of an imine since the latter was tentatively identified among the metabolites.

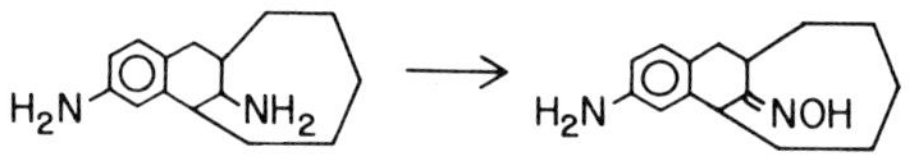

12 13

Oxidation to Nitroso-Compounds. C-Nitroso compounds containing α-hydrogens are isoelectronic with the more stable oximes to which they generally isomerize (scheme 3).

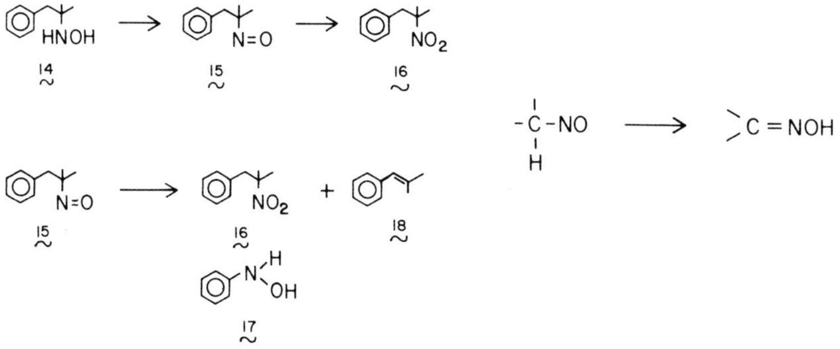

N-Hydroxyphentermine *(14)* (2-methyl-1-phenyl-2-propylhydroxylamine) [57] and phenylhydroxylamine *(17)* do not have α-hydrogens and thus form stable nitro-derivatives. 2-Methyl-2-nitroso-1-phenylpropane *(15)*, the nitroso-analog of phentermine, was originally reported as a metabolite of phentermine by Beckett and Belanger [4] in studies with rabbit liver homogenates. This nitroso-compound is also formed from *N*-hydroxyphentermine by rabbit liver microsomes [63] (scheme 3).

Very little, if any, of nitroso-compound *(15)* was noted in incubates of rat tissue [63]. Incubation of the nitroso-compound *(15)* with rabbit liver microsomes yielded a mixture of the nitro-compound *(16)* and dimethylstyrene *(18)* [63]. In subsequent experiments with reconstituted cytochrome P-450 systems [22] using isozymes from phenobarbital-induced rat, evidence was obtained for the indirect oxidation of *N*-hydroxyphentermine. *N*-hydroxyphentermine appears to uncouple the cytochrome P-450 system to reduce oxygen to superoxide which then oxidizes the hydroxylamine to the nitroso-state. This conversion has also been demonstrated in chemical systems [26]. The nitroso-function is rapidly oxidized to the nitro-state by oxygen.

Aromatic hydroxylamines are also oxidized to the nitroso-state [51]. Phenylhydroxylamine, in the presence of oxygen and ferrohemoglobin, is converted to nitrosobenzene. The hemoglobin is oxidized to ferrihemoglobin during this transformation, accounting for the methemoglobinemia associated with aromatic hydroxylamines. Proposed mechanisms for methemoglobinemia are described in chapter VI.

Evidence has also been reported for a one-electron oxidation of aromatic hydroxylamines to nitroxide radicals [88]. In this in vitro study with liver microsomes, the formation of free radicals was monitored by ESR techniques. The nitroxide radical, characterized by the spectral location of its ESR signal, was identified when a number of different aromatic hydroxylamines were present. The carcinogenic aromatic nitrogen compounds had a greater tendency to generate radical signals.

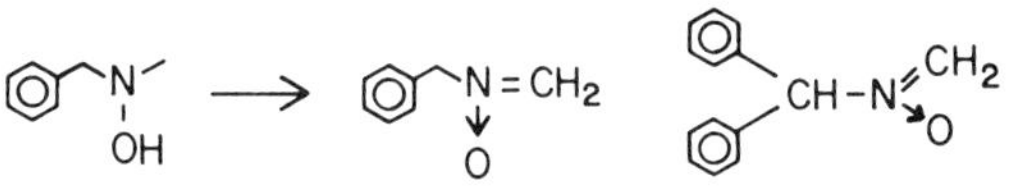

19 20 21

Oxidation to Nitrones. Nitrones, the *N*-oxidation products of secondary hydroxylamines, have also been identified as metabolites. The nitrone of *N*-methyl-*N*-benzylhydroxylamine *(20)* and the nitrone of *N*-methyl-*N*-benzhydrylhydroxylamine *(21)* were identified in liver microsomal incubates of the corresponding hydroxylamines [45]. The nitrone is indicated as a relatively unstable species which can undergo hydrolysis to the corresponding hydroxylamine and carbonyl compound [45], but more recent evidence suggests that their stability is highly dependent on structure. If decomposition does occur and is not recognized, the hydrolysis products of the nitrone can be interpreted as arising from *C*-oxidation and elimination. A nitrone of *N*-methylamphetamine has been reported and proposed to be the intermediate in the conversion of *N*-methylamphetamine to *N*-hydroxyamphetamine [14, 58].

After *N*-oxidation to the secondary hydroxylamine, a second two-electron oxidation of *N*-hydroxy-*N*-methylamphetamine will yield the nitrone *(22)* which can hydrolyze to formaldehyde and *N*-hydroxyamphetamine *(8)*. The latter compound can then be oxidized to the oxime [14, 62]. This hypothesis was validated by Lindeke et al. [58], who compared the rates of demethylation with rates of cytochrome P-450 MI complex (chapter III) formation for *N*-hydroxy-*N*-methylamphetamine and for *N*-methylamphetamine. The relationship between rate of formaldehyde and MI complex formation from *N*-methylamphetamine and the rapid rate of MI complex formation from *N*-hydroxy-*N*-methylamphetamine indicated that the nitrone eliminates formaldehyde to give the hydroxylamine product which is then converted to the MI complex:

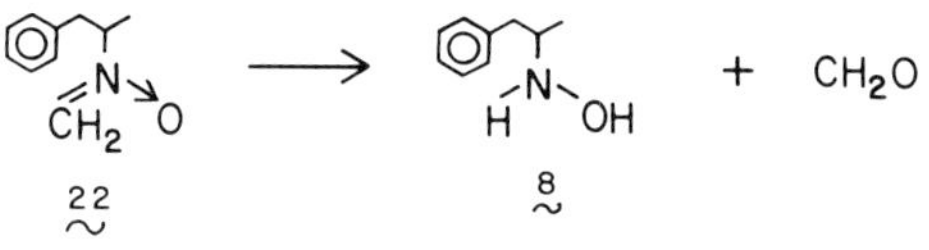

In a more recent investigation of the metabolism of methamphetamine, Yamada et al., [104] compared flavin monooxygenase and cytochrome-P-450-dependent formaldehyde generation by liver microsomes from rat and guinea pig. Demethylation by the flavin enzyme takes place through an initial *N*-hydroxylation pathway. These workers noted that the flavin pathway is favored in guinea pig tissue and both pathways are equally important in the rat.

Conjugation

The *O*-glucuronides of aliphatic [3, 8, 9] and aromatic hydroxylamines [40] are terminal metabolites and have been identified as urinary excretion products. The *N*-glucuronides of some arylhydroxylamines are thought to contribute to their carcinogenicity. The *N*-glucuronides are the major conjugates of these hydroxylamines, but whether they are formed preferentially to *O*-glucuronides or are more stable than the *O*-glucuronide is not known [46]. The *N*-glucuronides decompose in dilute acid to the hydroxylamines which, after protonation, can lose water to form the reactive nitrenium ion (see chapter VI).

This electrophilic species has been proposed as a reactive group that forms covalent bonds with tissue macromolecules. Another possible reactive group is a free radical [88]. The *O*-sulfate has also been proposed as a precursor to a reactive intermediate for carcinogenic arylhydroxylamines [18]. By analogy with the protonated hydroxylamine, the sulfate ester *(23)* readily loses the elements of sulfate to generate the nitrenium ion *(24)* which can react with tissue nucleophiles to form covalent bonds. However, Frederick et al. [25] have also suggested that a direct reaction can occur without conjugation. The subject of covalent bond formation is discussed in more details in chapter VI (reaction 7).

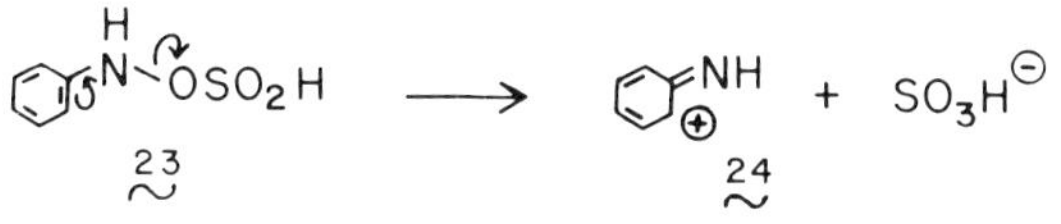

In the limited studies that have been conducted with arylalkylhydroxylamines, sulfate conjugates do not appear to be major excretory metabolites. Thus, most of the conjugates of *N*-hydroxy-4-chlorophentermine in urine were hydrolyzed by glucuronidase [9] indicating little, if any, sulfate ester was present. *N*-Hydroxyphentermine is also excreted as a conjugate [3], but the conditions used for hydrolysis did not distinguish between glucuronide and sulfate.

Hydrazines

Oxidation

Hydrazine is an amino analog of hydroxylamine, is of metabolic interest because of the toxicity of its metabolites (chapter VI) [95]. Unsymmetrical and symmetrical hydrazines differ in their biochemical properties. Unsymmetrical alkylhydrazines form spectrally defined complexes with cytochrome P-450, through an O_2/NADPH-dependent mechanism [35]. The properties of the complex are described in chapter III. The complex forms only with 1,1-disubstituted hydrazines (e.g. *25*) and not the mono- or 1,2-disubstituted hydrazines. Thus, oxidation of the primary amino group to some reactive moiety is highly likely. The diazene *(26)* has been proposed [35] as a reactive function, but free radicals have also been described [86]. Other hydrazine derivatives with a free terminal -NH_2 such as phenelzine *(27)* [69] and isoniazid [68] *(28)*, have also been shown to form inhibitory complexes with cytochrome P-450.

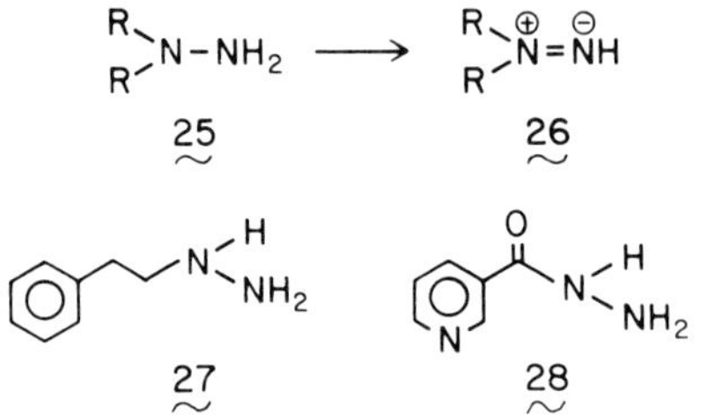

Although unsymmetrical hydrazines form inhibitory complexes with cytochrome P-450, they are still metabolized so that their metabolic transformation must be catalyzed by other enzymes. Prough et al. [77] have examined the oxidation of *N,N*-dialkylhydrazines by the NADPH-dependent FAD monooxygenase (chapter IV). The products of the reaction identified from 1,1-dimethylhydrazine *(29)* were formaldehyde and methylhydrazine *(30)*. The reaction path proposed to account for these metabolites is shown below (scheme 4):

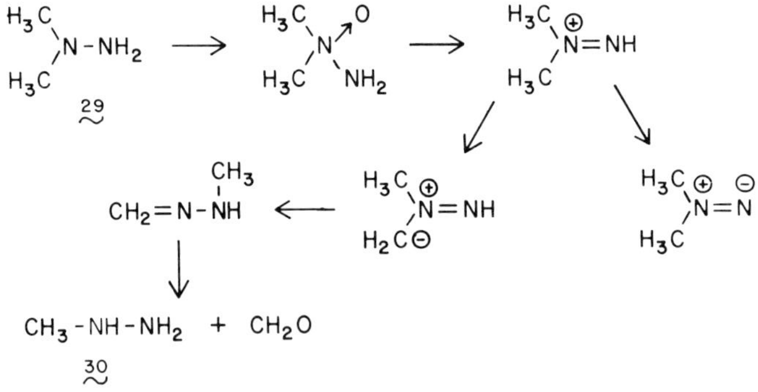

In separate incubations, methylhydrazine *(30)* was converted to methane. Monoalkylhydrazines are thought to be metabolized by oxidation to alkyldiazines which then decompose to the alkane and nitrogen [78]. Thus, methylhydrazine *(30)* could undergo *N*-oxidation to a diazine,

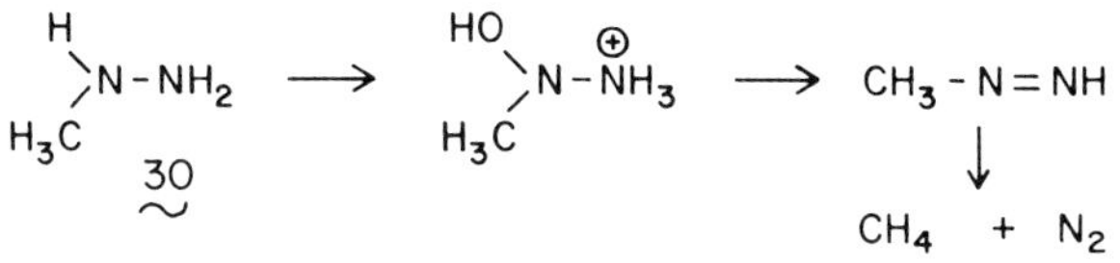

30

which in turn would generate methane by the reaction sequence indicated.

This pathway would also account for the formation of methane [19] during the metabolism of 1,1-dimethylhydrazine. These products of *N,N*-dimethylhydrazine metabolism could also be accounted for by carbon hydroxylation, but since the flavin adenine dinucleotide (FAD) monooxygenase only oxygenates nitrogen and sulfur atoms the pathways above are more likely.

The symmetrical dialkylhydrazines are oxidized by conversion to azocompounds which can then undergo a series of reactions that have been described for procarbazine *(31)* [101], and shown below (scheme 5). Except for bracketed structures, the compounds have been identified by mass spectral comparisons with authentic materials. The diazoalkane

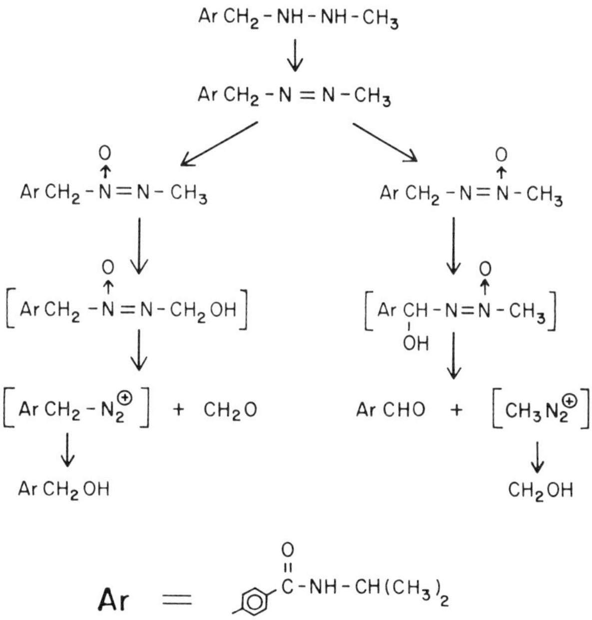

31

R-N_2^+, is a highly reactive alkylating agent and is proposed to be the species responsible for the cytotoxic actions of this anticancer drug.

An analogous series of reactions have been proposed on the basis of metabolites isolated from the urine of animals treated with 1,2-dimethylhydrazine [24] (scheme 6):

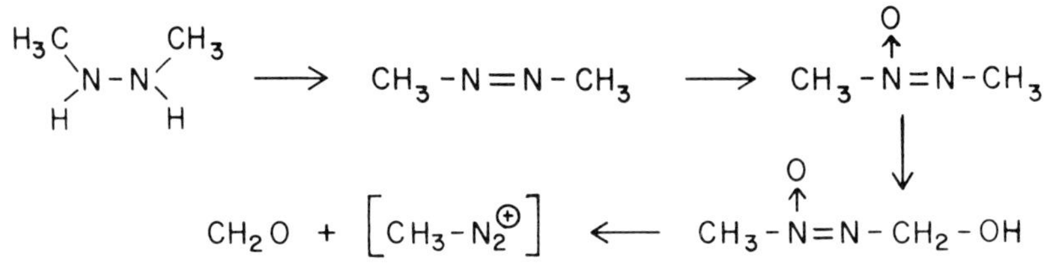

In other studies of hydrazines, the carbon moieties of acetylhydrazine and isopropylhydrazine were shown to form covalent bonds with microsomal protein [71]. The key intermediate proposed as the acylating or alkylating species was the diazine derivative formed via *N*-hydroxylation (scheme 7):

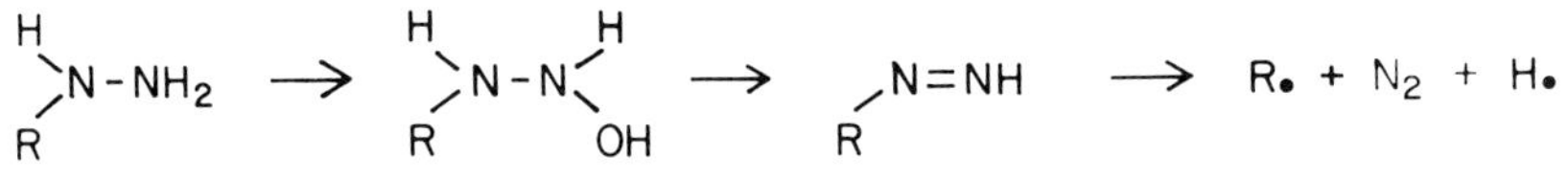

$R = CH_3\text{-}CO\text{—}, \quad (CH_3)_2CH\text{—}$

Hydrazines and hydrazides such as phenelzine *(27)* and isoniazid *(28)* are irreversible inhibitors of monoamine oxidase, and nitrogen oxidation is thought to be a key step in forming the inhibiting chemical species [50].

Hydrazine functions are attached to unsaturated carbon in two well-studied drugs, isoniazid *(28)* and hydralazine *(32)*.

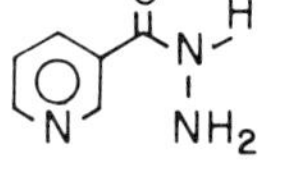

28

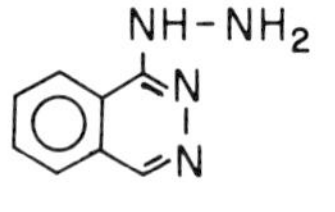

32

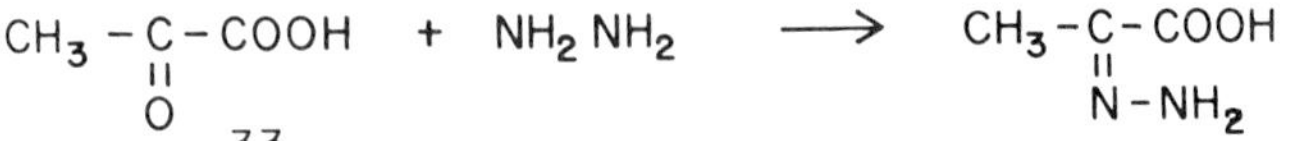

33 34

Both of these functions undergo acetylation and hydrolytic cleavage reactions to generate acetylhydrazine and hydrazine [72]. Hydrazine itself can react with endogenous carbonyl compounds, such as pyruvate *(33)* to form a hydrazone *(34)* [66].

These adducts have been identified in the plasma of patients treated with these drugs [72].

Conjugation

The hydrazine group is conjugated in the liver by acetylation. Recognition of the genetically transmitted polymorphism of the responsible enzymes has resulted in the characterization of two phenotypes in the general population, slow and fast acetylators. The slow acetylators are more inclined to the peripheral neuropathy associated with the toxicology of the antitubercular drug, 2-isonicotinyl-hydrazine (INH) [95, 100]. The responsible liver enzyme, *N*-acetyltransferase, also *N*-acetylates aromatic amines, such as the carcinogens 2-aminonaphthalene and 2-aminofluorene [30]. A limited structure-activity study suggests that lipophilic amines, particularly those containing planar aromatic rings, are good substrates for the *N*-acetyltransaminase.

Oximes

Oximes have been reported to undergo both oxidative and reductive metabolism. Coutts et al. [15] have reported oxidation of phenylacetone oxime to the corresponding nitro-compound in incubations of 12,000-*g* supernatants of rat liver homogenate. Other studies with the same oxide [62] have shown that it is stable in microsomal incubation mixtures so that the oxidation must be associated with the cytosolic component of the liver homogenate. *O*-Alkylation may stabilize oximes to metabolism as evidenced by fluvoxamine. Fluvoxamine is an *O*-substituted oxime, and urinary excretion studies with several species indicate that this oxime function is resistant to further metabolism. The compound is metabolized predominately at a terminal methoxyl function [84]. The oxidative metabolism of this class of compound is not well documented.

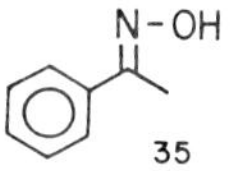

35

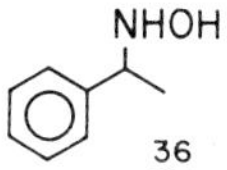

36

Acetophenone oxime *(35)* is reduced to the corresponding hydroxylamine *(36)* by microsomes fortified with NADPH. The reduction is oxygen-sensitive and carbon-monoxide-insensitive and exhibits substantial species differences. Rat liver homogenate reduces the oxime only to the level of the hydroxylamine, whereas homogenates from guinea pig, mouse and hamster liver reduce it to the -NH_2 state [34].

Nitroso-Compounds

As indicated earlier, aliphatic carbon nitroso-compounds containing an α-hydrogen are unstable with respect to the oxime. However, two compounds that do form a stable nitroso-state are phentermine *(37)* and amantadine *(38)* because they both lack α-hydrogens and cannot form oximes. The formation of the nitroso-metabolite has been reported for phentermine [4], *N*-hydroxyphentermine [3, 63], and amantadine [6].

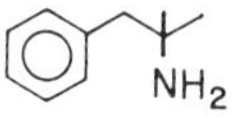

37

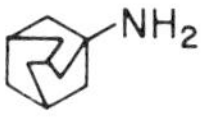

38

The nitroso analog of amantadine is reduced by rabbit liver microsomes by an NADPH- or NADH-dependent enzyme system that is not sensitive to oxygen or carbon monoxide [5]. Aromatic nitroso-compounds usually undergo reduction to the hydroxylamines so that hydroxylamine-nitroso-states appear to be rapidly reversible. In addition, however, aromatic nitroso-compounds condense to form bis-azoxy derivatives [54], and this chemical reaction could lead to the excreted metabolites.

Azo-Compounds

Aromatic azo-compounds are reduced to hydrazo-compounds, then cleaved to amines [27] by both cytochrome P-450 and NADPH cytochrome *c* reductase (scheme 8). This transformation is common for aro-

$$ArN{=}NAr' \longrightarrow \underset{H}{ArN}\text{-}\underset{H}{N}Ar' \longrightarrow ArNH_2 + Ar'NH_2$$

matic azo-compounds, suggesting that reduction is favored over oxidative reactions. Azo-reduction, catalyzed by enzymes of intestinal flora, is an important reaction in this tissue [31].

Nitro-Compounds

Reduction of Aromatic Nitro-Groups

Aromatic nitro-compounds are usually reduced to amino-compounds as evidenced by reports of the reduction of *p*-nitrobenzoate [28, 48], nitrotoluenes [81], nitrofurantoin [61], niridazole, a nitrothiazole [23], and chloramphenicol [29]. The enzymes responsible for the reduction of these compounds can be oxygen-sensitive or insensitive and are associated with both the endoplasmic reticulum and the cytosol. The apparent sensitivity to oxygen may in fact be the net result of an oxygen-dependent reoxidation of the nitroanion radical concomitantly generating superoxide [75]. Two enzymes of the endoplasmic reticulum, FAD-cytochrome *c* reductase [23], and cytochrome P-450 [28], reduce aromatic nitro-compounds by an oxygen-sensitive mechanism. The cytosolic enzymes are NADH-dependent and have been shown to carry out the reduction in a stepwise manner from nitro- to hydroxylamine to amine stage [48]. Nitrotoluenes appear to be metabolized by oxidation at the methyl carbon rather than by reduction of the nitro-group [17].

Chloramphenicol is another aromatic nitro-compound of interest because of its clinical importance and toxicity [55, 67]. One of the proposed intermediates in chloramphenicol toxicity is the nitroso-analog [105]. However, Ascherl et al. [1] have examined the metabolic disposition of the nitroso-analog of chloramphenicol and found that it is rapidly converted to the *N*-hydroxy derivative, its glutathione conjugate and the amine. Thus, the nitroso-compound appears to have a short half-life in the liver and under normal circumstances may not be present in significant levels. The hydroxylamine is much more stable.

Reduction of Nitrofurans

The chemistry of the nitrofurans differs from that of nitrobenzenes [39] in that reduction leads to a ring opening reaction which is thought to participate in the toxic action of the parent compound. These compounds are reduced by a variety of enzymes, such as xanthine oxidase [92], cytochrome *c* reductase [99], and undefined enzymes present in *E. coli* [75], and in microsomes from liver and kidney [106]. The reduction proceeds through the nitroso- and hydroxylamine to the amino state. Metabolites with these oxidation states have been demonstrated in in vivo metabolic studies. In those instances where products have been identified, hydroxylamines [23], amines [92], and nitriles [10, 99, 106] have been reported.

The proposed route(s) of formation of the identified products involve nitrogen in either the hydroxylamine or amine state. For example, the hydroxylamine state could account for an acrylonitrile metabolite identified by Zenser et al. [106] by the reactions proposed by Peterson et al. [75]:

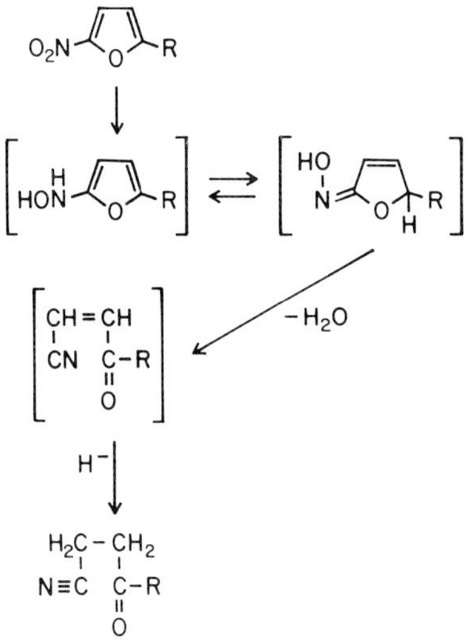

The reduction to the hydroxylamine state has been proposed to proceed by a sequential addition of electrons, yielding the nitro-anion radical, the nitroso- and then the hydroxylamine [36, 75]. The acrylonitrile is a reactive species that would readily add nucleophilic functions present in tissue. The formation of covalent bonds by such reactions is thought to be the initial step by which many compounds exert their toxicology response.

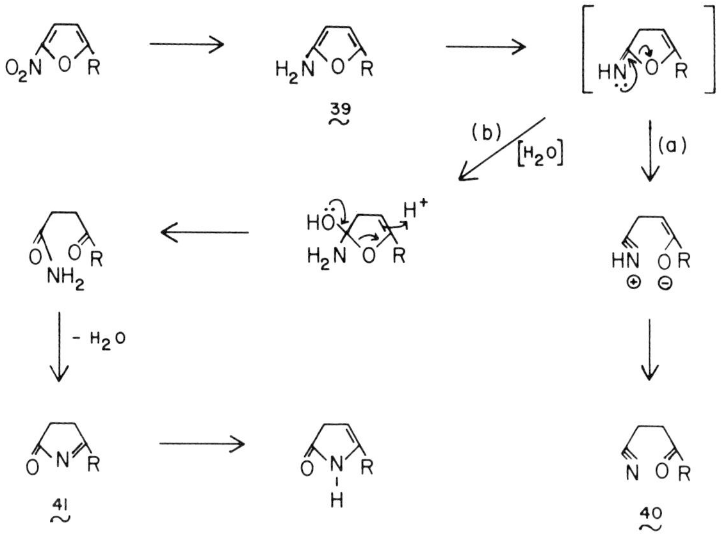

Reduction of nitrofuran to the amino state accounts for the other metabolites reported [92] according to the reaction shown above (scheme 9). The tautomer of the aminofuran *(39)* can undergo ring opening directly: (a) to yield the saturated nitrile *(40)*; or (b) to be hydrolyzed to the amide which can undergo ring closure to the pyrrolidone *(41)*.

Aliphatic Nitro-Compounds

There have been no reports of reduction of aliphatic nitro-groups by liver fractions. The effect of oxygen exclusion was examined in a study of denitrification [47], and under these conditions, liver microsomes did not significantly reduce 2-nitro-1-phenylpropane to the corresponding amine. Similar studies on the denitrification of 2-nitropropane indicate that this aliphatic compound is also not reduced by microsomes [98]. Further evidence to support the notion that aliphatic nitro-groups are not reduced is their inability to form metabolic intermediary complexes with NADPH and liver microsomes (chapter III). Some aliphatic nitro-compounds do form MI complexes in the presence of dithionite and liver microsomes [60]. Dithionite reduces the nitro-group to either the nitroso or hydroxylamino state, and either of these functions can be converted to ligands for reduced cytochrome P-450. If biochemical reduction of the nitro-group were possible, aliphatic nitro-compounds should be converted to P-450 ligands without prior chemical reduction.

Aliphatic nitro-compounds appear to interact with cytochrome P-450 by oxidation of the α-carbon atom to a carbonyl compound. Thus, aliphatic nitro-compounds with α-hydrogens such as 2-nitropropane [60] and 2-nitro-1-phenylpropane [42, 47] are converted to the corresponding ketones. This transformation can be rationalized as an α-carbon hydroxylation as indicated below (scheme 13). In the case of 2-nitropropane [98], the other product of the reaction has been shown to be NO_2^-:

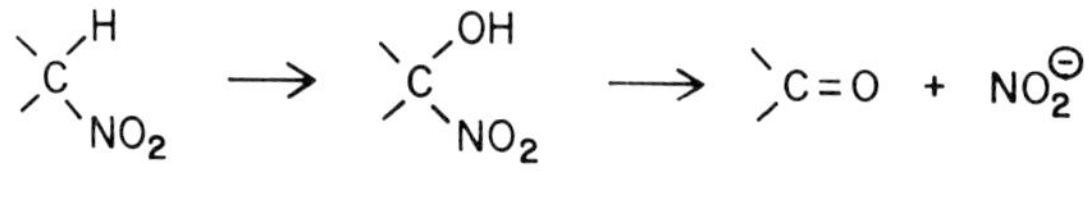

A recent study [93] investigating the reduction of aliphatic nitro-compounds by xanthine oxidase and by liver microsomes demonstrated that only conjugated aliphatic nitro-compounds are reduced by these systems. Thus, a series of saturated and unsaturated phenylalkylnitro-compounds were incubated with xanthine oxidase and hypoxanthine in the absence of oxygen. Only those nitro-compounds with a conjugated α,β-unsaturation *(42, 43)* were reduced *(42–46)*. The product in one case, however, was the dimer *(48)*.

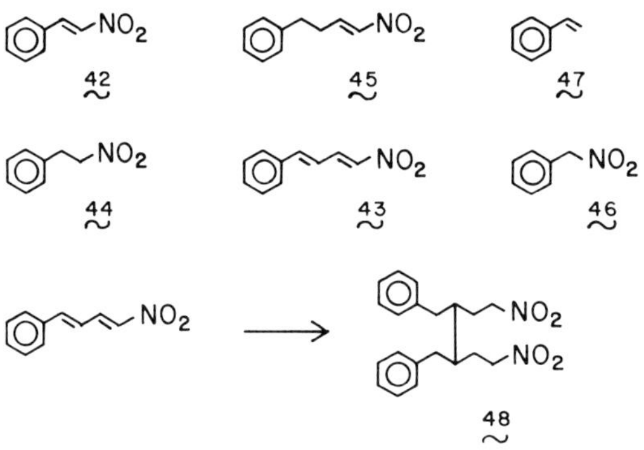

This reaction requires a conjugated unsaturated nitro-compound since the nitro-compounds, *44–46*, and styrene, *47*, were incapable of undergoing the reaction. This reductive dimerization of *42* was also demonstrated in incubates of rat liver microsomes or 105,000-*g* supernatant in the presence of NADPH, so that the reaction is limited to xanthine oxidase generated reducing equivalents.

The metabolism of five-membered heterocycles containing a nitro-group adjacent to a heteroatom has recently been reviewed as part of a symposium on the bioreduction of drugs. Miwa et al. [65] describe the detailed metabolic pathways involved in the conversion of ronidazole ([1-methyl-5-nitroimidazole-2yl]-methyl carbamate), an antiprotozoal agent. These studies utilized a series of isotopically substituted compounds to systematically determine the individual steps in covalent bond formation by this compound. Bigalow et al. [7] present a more general description of the biochemistry of the reduction of the nitro group in these substances. In another study of 5-nitroimidazole metabolism, tinidazole has been shown to undergo a rearrangement in which the 5-nitro group migrates to the 4 position during metabolism [102]. Superficially the reaction resembles the NIH, shift but mechanistic details have yet to be elucidated.

Conclusions

The metabolism of nitrogen functionalities reflects their chemistry, and it is important to recognize that this involves more than the oxidation state of the nitrogen. The extensive ring opening reactions and rearrangements that occur with the nitrofurans result from the oxidation state of the carbon atom to which the nitro group is attached. There is considerable difference between aromatic and aliphatic amines. The amino-group is attached to a trigonal carbon in aromatic amines and to a tetrahedral carbon in aliphatic amines. In addition to the steric effects of this difference, the delocalization of the unshared electrons on nitrogen with the aromatic system reduces the base strength and the nucleophilicity of the amino-group so that aromatic amines are essentially neutral at physiological pH. Aliphatic amines, on the other hand, are essentially charged cations. Since most of the metabolic transformations appear to involve neutral molecules, there is a great difference between aliphatic and aromatic amines in the proportion of reactive species available for enzymatic action at physiological pH.

The effects of small changes in structure on metabolic disposition have not been extensively studied. There are logistical reasons for the paucity of such studies in that for each potential substrate a series of metabolites must be prepared for use as analytical standards. In addition, separate analytical procedures must be developed for nonvolatile or unstable compounds. The latter problem can be reduced by use of liquid chromatography, and for this reason the technique has become extremely important in such studies [25].

When comparisons have been made between closely related substrates, selectivity, indicated by enantiomeric preferences or steric hindrance sensitivities by cytochrome P-450, are recognized. The basis for this discriminating ability of drug-metabolizing enzymes must be elucidated in order to predict pathways of drug metabolism, and no doubt will be the focus of drug metabolism research in the next few years.

References

1 Ascherl, M.; Eyer, P.; Kampffmeyer, H.: Formation and disposition of nitrosochloramphenicol in rat liver. Biochem. Pharmac. *34:* 3755–3763 (1985).

2 Beckett, A.H.; Al Sarraj, S.: The mechanism of oxidation of amphetamine enantiomorphs by liver microsomal preparation from different species. J. Pharm. Pharmac. *24:* 174–176 (1972).

3 Beckett, A.H.; Belanger, P.M.: The disposition of phentermine and its *N*-oxidized metabolic products in the rabbit. Xenobiotica *8:* 55–60 (1978).

4 Beckett, A.H.; Belanger, P.M.: The identification of three new metabolic products of phentermine after liver microsomal incubations. Xenobiotica *4:* 509–519 (1974).

5 Belanger, P.M.; Grech-Belanger, O.: Metabolic reduction of 1-nitrosoadamantadine by rabbit liver microsomes properties of a *C*-nitroso reductase system. Drug Metab. Disposit. *7:* 171–175 (1979).

6 Belanger, P.M.; Grech-Belanger, O.: The identification of the *N*-oxidized metabolic products of amantadine in the rabbit. Can. J. Pharm. Sci. *12:* 99–102 (1977).

7 Biaglow, J.E.; Varnes, M.E.; Roizen-Towle, L.; Clark, E.P.; Epp, E.R.; Astor, M.B.; Hall, E.J.: Biochemistry of reduction of nitro heterocycles. Biochem. Pharmac. *35:* 77–90 (1986).

8 Caldwell, J.: Conjugation reactions of nitrogen centers; in Jakoby, Bend, Caldwell, Metabolic basis of detoxication, pp. 292–306 (Academic Press, New York 1982).

9 Caldwell, J.; Koster, U.; Smith, R.L.; Williams, R.T.: Species variations in the *N*-oxidation of chlorphentermine. Biochem. Pharmac. *24:* 2225–2232 (1975).

10 Chatfield, D.H.: Disposition and metabolism of some nitrofurylthiazoles possessing antiparasitic activity. Xenobiotica *6:* 509–520 (1976).

11 Cho, A.K.; Lindeke, B.; Jenden, D.J.: Gas chromatography/mass spectrometry in the distribution and metabolism of phentermine; in Frigerio, Castagnoli, Mass spectrometry in biochemistry and medicine, pp. 83–90 (Raven Press, New York 1974).

12 Cho, A.K.; Lindeke, B.; Sum, C.Y.: The N-hydroxylation of phentermine (2-methyl-1-phenylisopropylamine): properties of the enzyme system. Drug Metab. Disposit. *2:* 1–8 (1974).

13 Coutts, R.T.; Beckett, A.H.: Metabolic *N*-oxidation of primary and secondary aliphatic medicinal amines. Drug Metab. Rev. *6:* 51–104 (1977).

14 Coutts, R.T.; Kovach, S.H.: Metabolism in vitro of *N*-methylamphetamine with rat liver homogenates. Biochem. Pharmac. *26:* 1043–1049 (1977).
15 Coutts, R.T.; Dawe, R.; Dawson, G.W.; Kovach, S.H.: In vitro metabolism of 1-phenyl-2-propanone oxime in rat liver homogenates. Drug Metab. Disposit. *4:* 35–39 (1976).
16 Cramer, J.W.; Miller, J.A.; Miller, E.C.: *N*-hydroxylation: a new metabolic reaction observed in the rat with the carcinogen 2-acetylaminofluorene. J. biol. Chem. *235:* 885–888 (1960).
17 De Bethizy, J.D.; Rickert, D.E.: Metabolism of nitrotoluenes by freshly isolated Fischer 344 rat hepatocytes. Drug Metab. Disposit. *12:* 45–50 (1984).
18 Debaun, J.R.; Miller, E.C.; Miller, J.A.: *N*-Hydroxy-2-acetylaminofluorene sulfotransferase: its probable role in carcinogenesis and protein-(methion-*S*-yl) binding in rat liver. Cancer Res. *30:* 577–595 (1979).
19 Dost, F.N.; Reed, D.J.: Methane formation in vivo from *N*-isopropyl α(2-methylhydrazine)-ρ-toluamide hydrochloride, a tumor inhibiting methylhydrazine derivative. Biochem. Pharmac. *16:* 1741–1746 (1967).
20 Duggan, D.E.; Baldwin, J.J.; Arison, B.H.; Rhodes, R.E.: *N*-glucoside formation as a detoxification mechanism in mammals. J. Pharmac. Exp. Ther. *190:* 563–569 (1974).
21 Duncan, J.D.; Cho, A.K.: *N*-Oxidation of phentermine to *N*-hydroxyphentermine by a reconstituted cytochrome P-450 oxidase system from rabbit liver. Molec. Pharmacol. *22:* 235–238 (1982).
22 Duncan, J.D.; Di Stefano, E.W.; Miwa, G.T.; Cho, A.K.: Role of superoxide in the *N*-oxidation of 2-methyl-1-phenyl-2-propylhydroxylamine by rat liver cytochrome P-450 system. Biochemistry *24:* 4155–4161 (1985).
23 Feller, D.R.; Morita, M.; Gillette, J.R.: Enzymatic reduction of niridazole by rat liver microsomes. Biochem. Pharmac. *20:* 203–215 (1971).
24 Fiala, E.S.: Investigations into the metabolism and mode of action of the colon carcinogen 1,2-dimethylhydrazine. Cancer *36:* 2407–2412 (1975).
25 Frederick, C.B.; Mays, J.B.; Ziegler, D.M.; Guengerich, F.P.; Kadlubar, F.F.: Cytochrome P-450- and flavin-containing monooxygenase-catalyzed formation of the carcinogen *N*-hydroxy-2-aminofluorene and its covalent binding the nuclear DNA. Cancer Res. *42:* 2671–2677 (1982).
26 Fukuto, J.M.; Di Stefano, E.W.; Burstyn, J.N.; Valentine, J.S.; Cho, A.K.: Mechanism of oxidation of *N*-hydroxyphentermine by superoxide. Biochemistry *24:* 4161–4167 (1985).
27 Gillette, J.R.: Reductive enzymes; in Brodie, Gillette, Concepts in biochemical pharmacology, part 2, pp. 349–361 (Springer, Berlin 1971).
28 Gillette, J.R.; Kamm, J.J.; Sasame, H.A.: Mechanism of *p*-nitrobenzoate reduction in liver. The possible role of cytochrome P-450 in liver microsomes. Molec. Pharmacol. *4:* 541–548 (1968).
29 Glasko, A.J.: Identification of chloramphenicol metabolites and some factors affecting metabolic disposition. Antimicrob. Agents Chemother. *1966:* 655–665.
30 Glowinski, I.B.; Radtke, H.E.; Weber, W.W.: Genetic variation in *N*-acetylation of carcinogenic arylamines by human and rabbit liver. Molec. Pharmacol. *14:* 940–949 (1978).
31 Goldman, P.: Role of the intestinal microflora; in Jakoby, Bend, Caldwell, Metabolic basis of detoxication, pp. 323–337 (Academic Press, New York 1982).

32 Goldstein, J.R.; Weaver, R.; Sundheimer, D.W.: Metabolism of 2-acetylaminofluorene by two 3-methylcholanthrene-inducible forms of rat liver cytochrome P-450. Cancer Res. *44:* 3768–3771 (1984).

33 Hallström, G.; Lindeke, B.; Khuthier, A.-H.; Al-Iraqi, A.L.: Microsomal formation and chemical decomposition of pargyline *N*-oxide. Chem.-Biol. Interact. *34.* 185–200 (1981).

34 Hes, J.; Sternson, L.A.: The metabolism of acetophenone oxime in rat liver homogenates. Drug Metab. Disposit. *2:* 345–350 (1974).

35 Hines, R.N.; Prough, R.A.: The characterization of an inhibitory complex formed with cytochrome P-450 and a metabolite of 1,1-disubstituted hydrazines. J. Pharmac. exp. Ther. *214:* 80–86 (1980).

36 Holtzman, J.L.: Role of reactive oxygen and metabolite binding in drug toxicity. Life Sci. *30:* 1–9 (1982).

37 Hucker, H.B.: Intermediates in drug metabolism. Drug Metab. Rev. *2:* 33–56 (1973).

38 Hucker, H.B.; Michniewicz, B.M.; Rhodes, R.E.: Phenylacetone oxime – an intermediate in the oxidative deamination of amphetamine. Biochem. Pharmac. *20:* 2123–2128 (1971).

39 Ichikawa, M.: Chemistry of nitrofurans; in Bryan, Carcinogenesis – a comprehensive survey, vol. 4, pp. 13–57 (Raven Press, New York 1978).

40 Irving, C.C.: Metabolic activation of *N*-hydroxy compounds by conjugation; in Bridges, Gorrod, Park, Biological oxidation of nitrogen, pp. 75–86 (Taylor & Francis, London 1972).

41 Iwasaki, K.; Noguchi, H.; Kato, R.; Imai, Y.; Sato, R.: Reduction of tertiary amine *N*-oxide by purified cytochrome P-450. Biochem. biophys. Res. Commun. *77:* 1143–1149 (1977).

42 Jonsson, J.; Kammerer, R.C.; Cho, A.K.: Metabolism of 2-nitro-1-phenylpropane to phenylacetone by rabbit liver microsomes. Res. Commun. chem. Pathol. Pharmacol. *18:* 75–82 (1977).

43 Jonsson, J.; Kammerer, R.C.; Hallström, G.; Lindeke, B.; Cho, A.K.: The metabolism of alpha-substituted phenethylamines by rabbit liver preparations; in Gorrod, Biological oxidation of nitrogen. 2nd Int. Symp. on the Biological Oxidation of Nitrogen in Organic Molecules, Chelsea, pp. 25–29 (Elsevier/North-Holland Biomedical Press, Amsterdam 1978).

44 Kadlubar, F.F.; Ziegler, D.M.: Properties of a NADH-dependent *N*-hydroxy amine reductase isolated from pig liver microsomes. Archs. Biochem. Biophys. *162:* 83–92 (1974).

45 Kadlubar, F.F.; McKee, E.M.; Ziegler, D.M.: Reduced pyridine nucleotide-dependent *N*-hydroxy amine oxidase and reductase activities of hepatic microsomes. Archs. Biochem. Biophys. *156:* 46–57 (1973).

46 Kadlubar, F.F.; Miller, J.A.; Miller, E.C.: Microsomal *N*-oxidation of the hepatocarcinogen *N*-methyl-4-aminoazobenzene and the reactivity of *N*-hydroxy-*N*-methyl-4-aminoazobenzene. Cancer Res. *36:* 1196–1206 (1977).

47 Kammerer, R.C.; Di Stefano, E.; Jonsson, J.; Cho, A.K.: Metabolism of 2-nitro-1-phenylpropane by rabbit liver microsomes. Biochem. Pharmac. *30:* 2257–2263 (1981).

48 Kato, R.; Oshima, T.; Takanaka, A.: Studies on the mechanism of nitro reduction by rat liver. Molec. Pharmacol. *5:* 487–498 (1969).

49 Kedderis, G.L.; Rickert, D.E.: Characterization of the oxidation of amine metabolites of nitrotoluenes by rat hepatic microsomes. Molec. Pharmacol. *28:* 207–214 (1985).

50 Kenny, W.C.; Naga, J.; Salach, J.; Singer, T.P.: Structure of the covalent phenylhydrazine adduct of monoamine oxidase: in Singer, von Korff, Murphy, Monoamine oxidase: structure, function and altered functions, pp. 25–37 (Academic Press, New York 1979).

51 Kiese, M.: The biochemical production of ferrihemoglobin-forming derivatives from aromatic amines and mechanisms of ferrihemoglobin formation. Pharmacol. Rev. *18:* 1091–1161 (1966).

52 King, C.M.; Allaben, W.T.: Arylhydroxamic acid acyltransferase; in Jakoby, Enzymatic basis of detoxication, vol. 2, pp. 187–197 (Academic Press, New York 1980).

53 Kloss, M.W.; Rosen, G.M.; Rauckman, E.J.: Cocaine mediated hepatotoxicity. A critical review. Biochem. Pharmac. *33:* 169–173 (1984).

54 Lenk, W.; Scharmer, U.: Reaction of microsomal and cytosolic enzymes with *N*-arylacetohydroxamic acids. Xenobiotica *10:* 573–591 (1980).

55 Lim, L.O.; Yunis, A.A.: Enzymatic reduction of chloramphenicol and nitrosochloramphenicol by rat liver microsomal preparations. Pharmacology *27:* 58–64 (1983).

56 Lindeke, B.; Cho, A.K.: *N*-dealkylation and deamination; in Jakoby, Bend, Caldwell, Metabolic basis of detoxication, pp. 105–126 (Academic Press, New York 1982).

57 Lindeke, B.; Cho, A.K.; Thomas, T.L.; Mechelson, L.: Microsomal *N*-hydroxylation of phenylalkylamines: identification of *N*-hydroxylated phenylalkylamines as their trimethylsilyl derivatives by GC/MS. Acta pharm. suec. *10:* 493–506 (1973).

58 Lindeke, B.; Paulsen, U.; Anderson, E.: Cytochrome P-450 complex formation in the metabolism of phenylalkylamines. IV. Spectral evidences for metabolic conversion of methamphetamine to *N*-hydroxyamphetamine. Biochem. Pharmac. *28:* 3629–3635 (1979).

59 Lotlikar, P.D.; Hong, Y.S.; Baldy, W.J.: Cytochrome P-450 dependent *N*-hydroxylation of 2-acetylaminofluorene; in Gorrod, Biological oxidation of nitrogen, pp. 185–193 (Elsevier/North-Holland Biomedical Press, Amsterdam 1978).

60 Mansuy, D.; Beaune, P.; Chottard, J.C.; Bertoli, J.F.; Gans, P.: The nature of the 455 nm absorbing complex formed during the cytochrome P-450-dependent oxidative metabolism of amphetamine. Biochem. Pharmac. *25:* 609–612 (1976).

61 Mason, R.P.; Holtzman, J.L.: The role of catalytic superoxide formation in the O_2 inhibition of nitroreductase. Biochem. biophys. Res. Commun. *67:* 1267–1280 (1975).

62 Matsumoto, R.M.; Cho, A.K.: Conversion of *N*-hydroxyamphetamine to phenylacetone oxime by rat liver microsomes. Biochem. Pharmac. *31:* 105–108 (1982).

63 Maynard, M.S.; Cho, A.K.: The metabolism of *N*-hydroxyphentermine (NOHP) in rabbit liver microsomes. Fed. Proc. *40:* 697 (1981).

64 McMahon, R.E.: Microsomal dealkylation of drugs: substrate specificity and mechanism. J. pharm. Sci. *55:* 457–466 (1966).

65 Miwa, G.T.; Wang, R.; Alvaro, R.; Walsh, J.S.; Lu, A.Y.H.: The metabolic activation of ronidazole [(1-methyl-5-nitroimidazole-2-yl)-methyl carbamate] to reactive metabolites by mammalian, cecal bacterial and *T. foetus* enzymes. Biochem. Pharmac. *35:* 33–36 (1986).

66 Moldeus, P.; Andersson, B.; Rahimtula, A.; Berggren, M.: Prostaglandin synthetase-catalyzed activation of paracetamol. Biochem. Pharmac. *31:* 1363–1368 (1982).
67 Morris, P.L.; Burke, T.R.; Pohl, L.R.: Reductive dechlorination of chloramphenicol by rat liver microsomes. Drug Metab. Disposit. *11:* 126–130 (1983).
68 Muakkassah, S.F.; Bidlack, W.R.; Yang, W.C.T.: Mechanism of the inhibitory action of isoniazid on microsomal drug metabolism. Biochem. Pharmac. *30:* 1651–1658 (1981).
69 Muakkassah, S.F.; Yang, W.C.T.: Mechanism of the inhibitory action of phenelzine on microsomal drug metabolism. J. Pharmac. exp. Ther. *219:* 147–155 (1981).
70 Nelson, S.D.; Dahlin D.C.; Rauckman, E.J.; Rosen, G.: Peroxidase-mediated formation of reactive metabolites of acetaminophen. Molec. Pharmacol. *20:* 195–199 (1981).
71 Nelson, S.D.; Mitchell, J.R.; Timbrell, J.A.; Snodgrass, W.R.; Corcoran, G.B.: Isoniazid and iproniazid activation of metabolites to toxic intermediate in man and rat. Science *193:* 901–902 (1976).
72 Noda, A.; Matsugame, K.; Yen, S.H.; Otsuji, N.; Iguchi, S.; Noda, H.: Identification of hydrazine derived from hydralazine in experimental animals. Chem. pharm. Bull. *27:* 1938–1941 (1979).
73 Parli, C.J.; McMahon, R.E.: The mechanism of microsomal deamination: heavy isotope studies. Drug Metab. Disposit. *1:* 337–340 (1973).
74 Parli, C.J.; Wang, N.; McMahon, R.E.: The enzymatic *N*-hydroxylation of an imine – a new cytochrome P-450-dependent reaction catalyzed by hepatic microsomal monooxygenase. J. biol. Chem. *246:* 6953–6955 (1971).
75 Peterson, F.J.; Mason, R.P.; Hovesepian, J.; Holtzman, J.L.: Oxygen-sensitive and -insensitive nitroreduction by *Escherichia coli* and rat hepatic microsomes. J. biol. Chem. *254:* 4009–4014 (1979).
76 Porter, C.C.; Arison, B.H.; Gruber, V.F.; Titus, D.C.; Vandenheuvel, W.J.A.: Human metabolism of cyproheptadine. Drug Metab. Disposit. *3:* 189–197 (1975).
77 Prough, R.A.; Freeman, P.C.; Hines, R.N.: The oxidation of hydrazine derivatives catalyzed by the purified liver microsomal FAD-containing monooxygenase. J. biol. Chem. *256:* 4178–4184 (1981).
78 Prough, R.A.; Wittkop, J.A.; Reed, D.J.: Evidence for the hepatic metabolism of some monoalkylhydrazines. Archs. Biochem. Biophys. *131:* 369–373 (1969).
79 Rauckman, E.J.; Rosen, G.M.; Kitchell, B.B.: Superoxide radical as an intermediate in the oxidation of hydroxylamines by mixed function amine oxidase. Molec. Pharmacol. *15:* 131–137 (1979).
80 Razzouk, C.; Mercier, M.; Roberfroid, M.: Biochemical basis for the resistance of guinea-pig and monkey to the carcinogenic effects of arylamines and arylamides. Xenobiotica *10:* 565–571 (1980).
81 Rickert, D.E.; Butterworth, B.E.; Popp, J.A.: Dinitrotoluene: acute toxicity, oncogenicity, genotoxicity and metabolism. CRC crit. Rev. Toxicol. *13:* 217–234 (1984).
82 Robertson, I.G.C.; Philpot, R.M.; Zieger, E.; Wolf, C.R.: Specificity of rabbit pulmonary cytochrome P-450 isozymes in the activation of several aromatic amines and aflatoxin B_1. Molec. Pharmacol. *20:* 662–668 (1981).
83 Robertson, I.G.C.; Serabjit-Singh, C.; Croft, J.E.; Philpot, R.M.: The relationship between increases in the hepatic content of cytochrome P-450, form 5 and in the

metabolism of aromatic amines to mutagenic products following treatment of rabbits with phenobarbital. Molec. Pharmacol. *24:* 156–162 (1983).

84 Ruijten, H.M.; De Bree, H.; Borst, A.J.M.; De Lange, N.; Scherpenisse, P.M.; Vincent, W.R.; Post, L.C.: Fluvoxamine: metabolic fate in animals. Drug Metab. Disposit. *12:* 82–92 (1984).

85 Santii, R.S.S.; Hopsu-Havu, V.K.: Transformylation of carcinogenic aromatic amines by kynurenine formamidase: a detoxication mechanism. Biochem. Pharmac. *17:* 1110–1113 (1968).

86 Sinha, B.K.; Motten, A.G.: Oxidative metabolism of hydralazine. Evidence for nitrogen centered radicals formation. Biochem. biophys. Res. Commun. *105:* 1044–1051 (1982).

87 Sisenwine, S.F.; Tio, C.O.: The metabolic disposition of dezocine in rhesus monkeys and female rats given ^{14}C-dezocine intragastrically. Drug Metab. Disposit. *9:* 37–42 (1981).

88 Stier, A.; Claus, R.; Lucke, A.; Reitz, I.: Redox cycle of stable mixed nitroxides formed from carcinogenic aromatic amines. Xenobiotica *10:* 661–673 (1980).

89 Sugiura, M.; Iwasaki, K.; Kato, R.: Reduction of tertiary amine *N*-oxides by liver microsomal cytochrome P-450. Molec. Pharmacol. *12:* 322–334 (1976).

90 Sum, C.Y., Cho, A.K.: Properties of microsomal enzyme systems that reduce *N*-hydroxyphentermine. Drug Metab. Disposit. *4:* 436–441 (1976).

91 Tang, B.K.; Kalow, W.; Grey, A.A.: Metabolic fate of phenobarbital in man: *N*-glucoside formation. Drug Metab. Disposit. *7:* 315–318 (1979).

92 Tatsumi, K.; Kitamura, S.; Yoshimura, H.: Reduction of nitrofuran derivatives by xanthine oxidase and microsomes. Archs. Biochem. Biophys. *175:* 131–137 (1976).

93 Tatsumi, K.; Yamada, H.; Yoshimura, H.; Kawazoe; Y.: Studies on enzymatic reduction of aliphatic nitro compounds: reductive dimerization of beta nitrostyrene and 1-nitro-4-phenylbutadiene. Archs. Biochem. Biophys. *213:* 689–694 (1982).

94 Thorgeirsson, S.S.; Jollow, D.J.; Sasame, H.A.; Green, I.; Mitchell, J.R.: The role of cytochrome P-450 in *N*-hydroxylation of 2-acetylaminofluorene. Molec. Pharmacol. *9:* 398–404 (1973).

95 Timbrell, J.A.: The role of metabolism in the hepatotoxicity of isoniazid and iproniazid. Drug Metab. Rev. *10:* 125–147 (1979).

96 Uehleke, H.: General biological aspects of *N*-hydroxylation. FEBS symp. *16:* 97–107 (1969).

97 Uetrecht, J.P.; Sweetman, B.J.; Woosley, R.L.; Oates, J.A.: Metabolism of procainamide to a hydroxylamine by rat and human hepatic microsomes. Drug Metab. Disposit. *12:* 77–81 (1984).

98 Ullrich, V.; Hermann, G.; Wever, P.: Nitrite formation from 2-nitropropane by microsomal monooxygenases. Biochem. Pharmac. *27:* 2301–2304 (1978).

99 Wang, C.Y.; Behrens, B.C.; Ichikawa, M.; Bryan, G.T.: Nitroreduction of 5-nitrofuran derivatives by rat liver xanthine oxidase and reduced nicotinamide adenine dinucleotide phosphate-cytochrome *c* reductase. Biochem. Pharmac. *23:* 3395–3404 (1974).

100 Weber, W.W.; Glowinski, I.B.: Acetylation; in Jakoby, Enzymatic basis of detoxication, vol. 2, pp. 169–186 (Academic Press, New York 1980).

101 Weinkam, R.J.; Shiba, D.A.: Metabolic activation of procarbazine. Life Sci. *22:* 937–946 (1978).

102 Wood, S.G.; Scott, P.W.; Chasseaud, L.F.; Faulkner, J.K.; Matthews, R.W.; Henrick, K.: A novel metabolite of tinidazole involving nitro-group migration. Xenobiotica *15:* 107–113 (1985).

103 Wright, J.; Cho, A.K.; Gal, J.: The role of *N*-hydroxyamphetamine in the metabolic deamination of amphetamine. Life Sci. *20:* 467–474 (1977).

104 Yamada, H.; Baba, T.; Hirata, Y.; Oguri, K.; Yoshimura, H.: Studies of *N*-demethylation of methamphetamine by liver microsomes of guinea pigs and rats: the role of flavin-containing monooxygenase and cytochrome P-450 systems. Xenobiotica *14:* 861–868 (1984).

105 Yunis, A.A.; Miller, A.M.; Salem, Z.; Arimura, G.K.: Nitroso-chloramphenicol. Possible mediator in chloramphenicol-induced aplastic anemia. J. Lab. clin. Med. *96:* 36–46 (1980).

106 Zenser, T.V.; Mattammal, M.B.; Palmier, M.O.; Davis, B.B.: Microsomal nitroreductase activity of rabbit kidney and bladder implication in 5-nitrofuran-induced toxicity. J. Pharmac. exp. Ther. *219:* 735–740 (1981).

A.K. Cho, PhD, UCLA Medical Center, Department of Pharmacology,
23–272 Center for the Health Sciences, 10833 Le Conte Avenue,
Los Angeles, CA 90024-1735 (USA)

Subject Index